T0256670

The Data Science Handbook

The Data Science Handbook

Field Cady

Registered Offices
John Wiley & Sons, Inc., 111 River Street, Hoboken, NJ 07030, USA

Editorial Office
111 River Street, Hoboken, NJ 07030, USA

For details of our global editorial offices, customer services, and more information about Wiley products visit us at www.wiley.com.

Wiley also publishes its books in a variety of electronic formats and by print-on-demand. Some content that appears in standard print versions of this book may not be available in other formats.

Library of Congress Cataloguing-in-Publication Data

Names: Cady, Field, 1984- author.
Title: The data science handbook / Field Cady.
Description: Hoboken, NJ: John Wiley & Sons, Inc., 2017. | Includes
  bibliographical references and index.
Identifiers: LCCN 2016043329 (print) | LCCN 2016046263 (ebook) | ISBN
  9781119092940 (cloth) | ISBN 9781119092933 (pdf) | ISBN 9781119092926
  (epub)
Subjects: LCSH: Databases--Handbooks, manuals, etc. | Statistics--Data
  processing--Handbooks, manuals, etc. | Big data--Handbooks, manuals, etc.
  | Information theory--Handbooks, manuals, etc.
Classification: LCC QA76.9.D32 C33 2017 (print) | LCC QA76.9.D32 (ebook) |
  DDC 005.74--dc23
LC record available at https://lccn.loc.gov/2016043329

Cover image: Сергей Хакимуллин/Gettyimages
Cover design by Wiley

Set in 10/12pt Warnock by SPi Global, Chennai, India

10 9 8 7 6 5 4 3 2 1

To my wife, Ryna. Thank you honey, for your support and for always believing in me.

Contents

Preface

This book was written to solve a problem. The people who I interview for data science jobs have sterling mathematical pedigrees, but most of them are unable to write a simple script that computes Fibonacci numbers (in case you aren't familiar with Fibonacci numbers, this takes about five lines of code). On the other side, employers tend to view data scientists as either mysterious wizards or used-car salesmen (and when data scientists can't be trusted to write a basic script, the latter impression has some merit!). These problems reflect a fundamental misunderstanding, by all parties, of what data science is (and isn't) and what skills its practitioners need.

When I first got into data science, I was part of that problem. Years of doing academic physics had trained me to solve problems in a way that was long on abstract theory but short on common sense or flexibility. Mercifully, I also knew how to code (thanks, Google™ internships!), and this let me limp along while I picked up the skills and mindsets that actually mattered.

Since leaving academia, I have done data science consulting for companies of every stripe. This includes web traffic analysis for tiny start-ups, manufacturing optimizations for Fortune 100 giants, and everything in between. The problems to solve are always unique, but the skills required to solve them are strikingly universal. They are an eclectic mix of computer programming, mathematics, and business savvy. They are rarely found together in one person, but in truth they can be learned by anybody.

A few interviews I have given stand out in my mind. The candidate was smart and knowledgeable, but the interview made it painfully clear that they were unprepared for the daily work of a data scientist. What do you do as an interviewer when the candidate starts apologizing for wasting your time? We ended up filling the hour with a crash course on what they were missing and how they could go out and fill the gaps in their knowledge. They went out, learned what they needed to, and are now successful data scientists.

I wrote this book in an attempt to help people like that out, by condensing data science's various skill sets into a single, coherent volume. It is hands-on and to the point: ideal for somebody who needs to come up to speed quickly or solve a problem on a tight deadline. The educational system has not yet caught up to the demands of this new and exciting field, and my hope is that this book will help you bridge the gap.

Field Cady
September 2016
Redmond, Washington

1

Introduction: Becoming a Unicorn

"Data science" is a very popular term these days, and it gets applied to so many things that its meaning has become very vague. So I'd like to start this book by giving you the definition that I use. I've found that this one gets right to the heart of what sets it apart from other disciplines. Here goes:

> Data science means doing analytics work that, for one reason or another, requires a substantial amount of software engineering skills.

Sometimes, the final deliverable is the kind of thing a statistician or business analyst might provide, but achieving that goal demands software skills that your typical analyst simply doesn't have. For example, a dataset might be so large that you need to use distributed computing to analyze it or so convoluted in its format that many lines of code are required to parse it. In many cases, data scientists also have to write big chunks of production software that implement their analytics ideas in real time. In practice, there are usually other differences as well. For example, data scientists usually have to extract features from raw data, which means that they tackle very open-ended problems such as how to quantify the "spamminess" of an e-mail.

It's very hard to find people who can construct good statistical models, hack quality software, and relate this all in a meaningful way to business problems. It's a lot of hats to wear! These individuals are so rare that recruiters often call them "unicorns."

The message of this book is that it is not only possible but also relatively straightforward to become a "unicorn." It's just a question of acquiring the particular balance of skills required. Very few educational programs teach all of those skills, which is why unicorns are rare, but that's mostly a historical accident. It is perfectly reasonable for a single person to have the whole palette of abilities, provided they're willing to ignore the traditional boundaries between different disciplines.

This book aims to teach you everything you'll need to know to be a competent data scientist. My guess is that you're either a computer programmer

The Data Science Handbook, First Edition. Field Cady.
© 2017 John Wiley & Sons, Inc. Published 2017 by John Wiley & Sons, Inc.

looking to learn about analytics or more of a mathematician trying to bone up on their coding. You might also be a businessperson who needs the technical skills to answer your business questions or simply an interested layman. Whoever you are though, this book will teach you the concepts you need.

This book is not comprehensive. Data science is too big an area for any person or book to cover all of it. Besides, the field is changing so fast that any "comprehensive" book would be out-of-date before it came off the presses. Instead, I have aimed for two goals. First, I want to give a solid grounding in the big picture of what data science is, how to go about doing it, and the foundational concepts that will stand the test of time. Second, I want to give a "complete" skill set, in the sense that you have the nuts-and-bolts knowledge to go out and do data science work (you can code in Python, you know the libraries to use, most of the big machine learning models, etc.), even if particular projects or companies might require that you pick up a new skill set from somewhere else.

1.1 Aren't Data Scientists Just Overpaid Statisticians?

Nate Silver, a statistician famous for accurate forecasting of US elections, once famously said: "I think data scientist is a sexed-up term for statistician." He has a point, but what he said is only partly true. The discipline of statistics deals mostly with rigorous mathematical methods for solving well-defined problems. Data scientists spend most of their time getting data into a form where statistical methods could even be applied. This involves making sure that the analytics problem is a good match to business objectives, extracting meaningful features from the raw data and coping with any pathologies of the data or weird edge cases. Once that heavy lifting is done, you can apply statistical tools to get the final results, although, in practice, you often don't even need them. Professional statisticians need to do a certain amount of preprocessing themselves, but there is a massive difference in degree.

Historically, data science emerged as a field independently from statistics. Most of the first data scientists were computer programmers or machine learning experts who were working on Big Data problems. They were analyzing datasets of the kind that statisticians don't touch: HTML pages, image files, e-mails, raw output logs of web servers, and so on. These datasets don't fit the mold of relational databases or statistical tools, so for decades, they were just piling up without being analyzed. Data science came into being as a way to finally milk them for insights.

In 20 years, I suspect that statistics, data science, and machine learning will blur into a single discipline. The differences between them are, after all, really just a matter of degree and/or historical accident. But in practical terms, for the time being, solving data science problems requires skills that a normal statistician does not have. In fact, these skills, which include extensive software engineering and domain-specific feature extraction, constitute the overwhelming majority of the work that needs to be done. In the daily work of a data scientist, statistics plays second fiddle.

1.2   How Is This Book Organized?

This book is organized into three sections. The first, The Stuff You'll Always Use, covers topics that, in my experience, you will end up using in almost any data science project. They are core skills, which are absolutely indispensable for data science at any level.

The first section was also written with an eye toward people who need data science to answer a specific question but do not aspire to become full-fledged data scientists. If you are in this camp, then there is a good chance that Part I of the book will give you everything you need.

The second section, Stuff You Still Need to Know, covers additional core skills for a data scientist. Some of these, such as clustering, are so common that they almost made it into the first section, and they could easily play a role in any project. Others, such as natural language processing, are somewhat specialized subjects that are critical in certain domains but superfluous in others. In my judgment, a data scientist should be conversant in all of these subjects, even if they don't always use them all.

The final section, Stuff That's Good to Know, covers a variety of topics that are optional. Some of these chapters are just expansions on topics from the first two sections, but they give more theoretical background and discuss some additional topics. Others are entirely new material, which does come up in data science, but which you could go through a career without ever running into.

1.3   How to Use This Book?

This book was written with three use cases in mind:

1) You can read it cover-to-cover. If you do that, it should give you a self-contained course in data science that will leave you ready to tackle real problems. If you have a strong background in computer programming, or in mathematics, then some of it will be review.
2) You can use it to come quickly up to speed on a specific subject. I have tried to make the different chapters pretty self-contained, especially the chapters after the first section.

3) The book contains a lot of sample codes, in pieces that are large enough to use as a starting point for your own projects.

1.4 Why Is It All in Python™, Anyway?

The example code in this book is all in Python, except for a few domain-specific languages such as SQL. My goal isn't to push you to use Python; there are lots of good tools out there, and you can use whichever ones you want.

However, I wanted to use one language for all of my examples. This keeps the book readable, and it also lets readers follow the whole book while only knowing one language. Of the various languages available, there are two reasons why I chose Python:

1) Python is the most popular language for data scientists. R is its only major competitor, at least when it comes to free tools. I have used both extensively, and I think that Python is flat-out better (except for some obscure statistics packages that have been written in R and that are rarely needed anyway).
2) I like to say that for any task, Python is the second-best language. It's a jack-of-all-trades. If you only need to worry about statistics, or numerical computation, or web parsing, then there are better options out there. But if you need to do all of these things within a single project, then Python is your best option. Since data science is so inherently multidisciplinary, this makes it a perfect fit.

As a note of advice, it is much better to be proficient in one language, to the point where you can reliably churn out code that is of high quality, than to be mediocre at several.

1.5 Example Code and Datasets

This book is rich in example code, in fairly long chunks. This was done for two reasons:

1) As a data scientist, you need to be able to read longish pieces of code. This is a nonoptional skill, and if you aren't used to it, then this will give you a chance to practice.
2) I wanted to make it easier for you to poach the code from this book, if you feel so inclined.

You can do whatever you want with the code, with or without attribution. I release it into the public domain in the hope that it can give some people a small leg up. You can find it on my GitHub page at www.github.com/field-cady.

The sample data that I used comes in two forms:

1) Test datasets that are built into Python's scientific libraries
2) Data that is pulled off the Internet, from sources such as Yahoo and Wikipedia. When I do this, the example scripts will include code that pulls the data.

1.6 Parting Words

It is my hope that this book not only teaches you how to do nut-and-bolts data science but also gives you a feel of how exciting this deeply interdisciplinary subject is. Please feel free to reach out to me at www.fieldcady.com or field. cady@gmail.com with comments, errata, or any other feedback.

Part 1

The Stuff You'll Always Use

The first section of this book covers core topics that everybody doing data science should know. This includes people who are not interested in being professional data scientists, but need to know just enough to solve some specific problem. These are the subjects that will likely arise in every data science project you do.

The Data Science Handbook, First Edition. Field Cady.
© 2017 John Wiley & Sons, Inc. Published 2017 by John Wiley & Sons, Inc.

2

The Data Science Road Map

In this chapter, I will give you a high-level overview of the process of data science. I will focus on the different stages of data science work, including common pain points, key things to get right, and where data science parts ways from other disciplines.

The process of solving a data science problem is summarized in the following figure, which I called the Data Science Road Map.

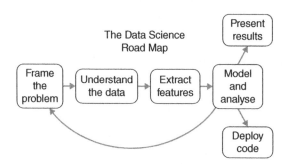

The first step is always to frame the problem: understand the business use case and craft a well-defined analytics problem (or problems) out of it. This is followed by an extensive stage of grappling with the data and the real-world things that it describes, so that we can extract meaningful features. Finally, these features are plugged into analytical tools that give us hard numerical results.

Before I go into more detail about the different stages of the roadmap, I want to point out two things.

The first is that "Model and Analyze" loops back to framing the problem. This is one of the key features of data science that differentiate it from traditional software engineering. Data scientists write code, and they use many of the same tools as software engineers. However, there is a tight feedback loop between data science work and the real world. Questions are always being reframed as new insights become available, and, as a result, data scientists

The Data Science Handbook, First Edition. Field Cady.
© 2017 John Wiley & Sons, Inc. Published 2017 by John Wiley & Sons, Inc.

must keep their code base extremely flexible and always have an eye toward the real-world problem they are solving. Ideally, you will follow the loop back many times, constantly refining your methods and producing new insights.

The second point is that there are two different (although not mutually exclusive) ways to exit the road map: presenting results and deploying code. My friend Michael Li, a data scientist who founded The Data Incubator, likened this to having two different types of clients: humans and machines. They require distinct skill sets and modifications to every stage of the data science road map.

If your clients are humans, then usually you are trying to use available data sources to answer some kind of business problem. Examples would be the following:

- Identifying leading indicators of spikes in the price of a stock, so that people can understand what causes price spikes
- Determining whether customers break down into natural subtypes and what characteristics each type has
- Assessing whether traffic to one website can be used to predict traffic to another site.

Typically, the final deliverable for work such as this will be a PowerPoint slide deck or a written report. The goal is to give business insights, and often these insights will be used for making key decisions. This kind of data science also functions as a way to test the waters and see whether some analytics approach is worth a larger follow-up project that may result in production software.

If your clients are machines, then you are doing something that blends into software engineering, where the deliverable is a piece of software that performs some analytics work. Examples would be the following:

- Implementing the algorithm that chooses which ad to show to a customer and training it on real data
- Writing a batch process that generates daily reports based on company records generated that day, using some kind of analytics to point out salient patterns

In these cases, your main deliverable is a piece of software. In addition to performing a useful task, it had better work well in terms of performance, robustness to bad inputs, and so on.

Once you understand who your clients are, the next step is to determine what you'll be doing for them. In the next section, I will show you how to do this all-important step.

2.1 Frame the Problem

The difference between great and mediocre data science is not about math or engineering: it is about asking the right question(s). Alternately, if you're trying to build some piece of software, you need to decide what exactly that software should do. No amount of technical competence or statistical rigor can make up for having solved a useless problem.

If your clients are humans, most projects start with some kind of extremely open-ended question. Perhaps there is a known pain point, but it's not clear what a solution would look like. If your clients are machines, then the business problem is usually pretty clear, but there can be a lot of ambiguity about what constraints there might be on the software (languages to use, runtime, how accurate predictions need to be, etc.). Before diving into actual work, it's important to clarify exactly what would constitute a solution to this problem. A "definition of done" is a good way to put it: what criteria constitute a completed project, and (most importantly) what would be required to make the project a success?

For large projects, these criteria are typically laid out in a document. Writing that document is a collaborative process involving a lot of back-and-forth with stakeholders, negotiation, and sometimes heated disagreement. In consulting, these documents are often called "statements of work" or SOWs. Within a company that is creating a product (as opposed to just a stand-alone investigation), they are often referred to as "project requirements documents" or PRDs.

The main purpose of an SOW is to get everybody on the same page about what exactly work should be done, what the priorities are, and what expectations are realistic. Business problems are typically very vague to start off with, and it takes a lot of time and effort to follow a course of action through to the final result. So before investing that effort, it is critical to make sure that you are working on the right problem.

There is, however, also an element of self-defense. Sometimes, it ends up being impossible to solve a problem with the available data. Or, the data is corrupted. Maybe clients decide that the project isn't important anymore. A good SOW makes it impossible for people to attack or sue you for wasting time and money on work that they now claim that they didn't want done.

Having an SOW doesn't set things in stone. There are course corrections based on preliminary discoveries. Sometimes, people change their minds after the SOW has been signed. It happens. But crafting an SOW is the best way to make sure that all efforts are pointed in the most useful direction.

2.2 Understand the Data: Basic Questions

Once you have access to the data you'll be using, it's good to have a battery of standard questions that you always ask about it. This is a good way to hit the ground running with your analyses, rather than risk analysis paralysis. It is also a good safeguard to identify problems with the data as quickly as possible.

A few good generic questions to ask are as follows:

- How big is the dataset?
- Is this the entire dataset?
- Is this data representative enough? For example, maybe data was only collected for a subset of users.
- Are there likely to be gross outliers or extraordinary sources of noise? For example, 99% of the traffic from a web server might be a single denial-of-service attack.
- Might there be artificial data inserted into the dataset? This happens a lot in industrial settings.
- Are there any fields that are unique identifiers? These are the fields you might use for joining between datasets, etc.
- Are the supposedly unique identifiers actually unique? What does it mean if they aren't?
- If there are two datasets A and B that need to be joined, what does it mean if something in A doesn't matching anything in B?
- When data entries are blank, where does that come from?
- How common are blank entries?

SOWs will often include an appendix that describes the available data. If any of them can't be answered up front, it is common to clear them up as a first round of analysis and make sure that everybody agrees that these answers are sane.

The most important question to ask about the data is whether it can solve the business problem that you are trying to tackle. If not, then you might need to look into additional sources of data or modify the work that you are planning.

Speaking from personal experience, I have been inclined to neglect these preliminary questions. I am excited to get into the actual analysis, so I've sometimes jumped right in without taking the time to make sure that I know what I'm doing. For example, I once had a project where there was a collection of motors and time series data monitoring their physical characteristics: one time series per motor. My job was to find leading indicators of failure, and I started doing this by comparing the last day's worth of time series for a given motor (i.e., the data taken right before it failed) against its previous data. Well, I realized a couple weeks in that sometimes the time series stopped long before the motor actually failed, and in other cases, the time series data continued long after the motor was dead. The actual times the motors had died were listed in a separate table, and it would have been easy for me to double-check early on that they corresponded to the ends of the time series.

2.3 Understand the Data: Data Wrangling

Data wrangling is the process of getting the data from its raw format into something suitable for more conventional analytics. This typically means creating a software pipeline that gets the data out of wherever it is stored, does any cleaning or filtering necessary, and puts it into a regular format.

Data wrangling is the main area where data scientists need skills that a traditional statistician or analyst doesn't have. The data is often stored in a special-purpose database that requires specialized tools to access. There could be so much of it that Big Data techniques are required to process it. You might need to use performance tricks to make things run quickly. Especially with messy data, the preprocessing pipelines are often so complex that it is very difficult to keep the code organized.

Speaking of messy data, I should tell you this upfront: industrial datasets are *always* more convoluted than you would think they reasonably should be. The question is not whether the problems exist but whether they impact your work. My recipe for figuring out how a particular dataset is broken includes the following:

1) If the raw data is text, look directly at the plain files in a text editor or something similar. Things such as irregular date formats, irregular capitalizations, and lines that are clearly junk will jump out at you.
2) If there is a tool that is supposed to be able to open or process the data, make sure that it can actually do it. For example, if you have a CSV file, try opening it in something that reads data frames. Did it read all the rows in? If not, maybe some rows have the wrong number of entries. Did the column that is supposed to be a datetime get read in as a datetime? If not, then maybe the formatting is irregular.
3) Do some histograms and scatterplots. Are these numbers realistic, given what you know about the real-life situation? Are there any massive outliers?
4) Take some simple questions that you already know the (maybe approximate) answer to, answer them based on this data, and see if the results agree. For example, you might try to calculate the number of customers by counting how many unique customer IDs there are. If these numbers don't agree, then you've probably misunderstood something about the data.

2.4   Understand the Data: Exploratory Analysis

Once you have the data digested into a usable format, the next step is exploratory analysis. This basically means poking around in the data, visualizing it in lots of different ways, trying out different ways to transform it, and seeing what there is to see. This stage is very creative, and it's a great place to let your curiosity run a little wild. Feel free to calculate some correlations and similar things, but don't break out the fancy machine learning classifiers. Keep things simple and intuitive.

There are two things that you typically get out of exploratory analysis:

1) You develop an intuitive feel for the data, including what the salient patterns look like visually. This is especially important if you're going to be working with similar data a lot in the future.
2) You get a list of concrete hypotheses about what's going on in the data. Oftentimes, a hypothesis will be motivated by a compelling graphic that you generated: a snapshot of a time series that shows an unmistakable pattern, a scatterplot demonstrating that two variables are related to each other, or a histogram that is clearly bimodal.

A common misconception is that data scientists don't need visualizations. This attitude is not only inaccurate: it is very dangerous. Most machine learning algorithms are not inherently visual, but it is very easy to misinterpret their outputs if you look only at the numbers; there is no substitute for the human eye when it comes to making intuitive sense of things.

2.5 Extract Features

This stage has a lot of overlap with exploratory analysis and data wrangling. A feature is really just a number or a category that is extracted from your data and describes some entity. For example, you might extract the average word length from a text document or the number of characters in the document. Or, if you have temperature measurements, you might extract the average temperature for a particular location.

In practical terms, feature extraction means taking your raw datasets and distilling them down into a table with rows and columns. This is called "tabular data." Each row corresponds to some real-world entity, and each column gives a single piece of information (generally a number) that describes that entity. Virtually all analytics techniques, from lowly scatterplots to fancy neural networks, operate on tabular data.

Extracting good features is the most important thing for getting your analysis to work. It is much more important than good machine learning classifiers, fancy statistical techniques, or elegant code. Especially if your data doesn't come with readily available features (as is the case with web pages, images, etc.), how you reduce it to numbers will make the difference between success and failure.

Feature extraction is also the most creative part of data science and the one most closely tied to domain expertise. Typically, a really good feature will correspond to some real-world phenomenon. Data scientists should work closely with domain experts and understand what these phenomena mean and how to distill them into numbers.

Sometimes, there is also room for creativity as to what entities you are extracting features about. For example, let's say that you have a bunch of transaction logs, each of which gives a person's name and e-mail address. Do you want to have one row per human or one row per e-mail address? For many real-world situations, you want one row per human (in which case, the number of unique e-mail addresses they have might be a good feature to extract!), but that opens the very thorny question of how you can tell when two people are the same based on their names.

Most features that we extract will be used to predict something. However, you may also need to extract the thing that you are predicting, which is also called the target variable. For example, I was once tasked with predicting whether my client's customers would lose their brand loyalty. There was no "loyalty" field in the data: it was just a log of various customer interactions and transactions. I had to figure out a way to measure "loyalty."

2.6 Model

Once features have been extracted, most data science projects involve some kind of machine learning model. Maybe this is a classifier that guesses whether a customer is still loyal, a regression model that predicts a stock's price on the next day, or a clustering algorithm that breaks customers into different segments.

In many data science projects, the modeling stage is quite simple: you just take a standard suite of models, plug your data into each one of them, and see which one works best. In other cases, a lot of care is taken to carefully tune a model and eek out every last bit of performance.

Really this should happen at every stage of a data science project, but it becomes especially crucial when analyzing the results of the modeling stage. If you have identified different clusters, what do they correspond to? Does your classifier work well enough to be useful? Is there anything interesting about the cases in which it fails?

This stage is what allows for course corrections in a project and gives ideas for what to do differently if there is another iteration.

If your client is a human, it is common to use a variety of models, tuned in different ways, to examine different aspects of your data. If your client is a machine though, you will probably need to zero in on a single, canonical model that will be used in production.

2.7 Present Results

If your client is a human, then you will probably have to give either a slide deck or a written report describing the work you did and what your results were. You are also likely to have to do this even if your main clients are machines.

Communication in slide decks and prose is a difficult, important skill set in itself. But it is especially tricky with data science, where the material you are communicating is highly technical and you are presenting to a broad audience. Data scientists must communicate fluidly with business stakeholders, domain experts, software engineers, and business analysts. These groups tend to have different knowledge bases coming in, different things they will be paying attention to, and different presentation styles to which they are accustomed.

2.8 Deploy Code

If your ultimate clients are computers, then it is your job to produce code that will be run regularly in the future by other people. Typically, this falls into one of two categories:

1) Batch analytics code. This will be used to redo an analysis similar to the one that has already been done, on data that will be collected in the future. Sometimes, it will produce some human-readable analytics reports. Other times, it will train a statistical model that will be referenced by other code.
2) Real-time code. This will typically be an analytical module in a larger software package, written in a high-performance programming language and adhering to all the best practices of software engineering.

There are three typical deliverables from this stage:

1) The code itself.
2) Some documentation of how to run the code. Sometimes, this is a stand-alone work document, often called a "run book." Other times, the documentation is embedded in the code itself.
3) Usually, you need some way to test code that ensures that your code operates correctly. Especially for real-time code, this will normally take the form of unit tests. For batch processes, it is sometimes a sample input dataset (designed to illustrate all the relevant edge cases) along with what the output should look like.

In deploying code, data scientists often take on a dual role as full-fledged software engineers. Especially with very intricate algorithms, it often just isn't practical to have one person spec it out and another implement the same thing for production.

2.9 Iterating

Data science is a deeply iterative process, even more so than typical software engineering. This is because in software, you generally at least know what you're ultimately aiming to create, even if you take an iterative approach to writing it. But in data science, it is usually an open question of what features will end up being useful to extract and what model you will train. For this reason, the data science process should be built around the goal of being able to change things painlessly.

My recommendations are as follows:

- Try to get preliminary results as quickly as possible after you've understood the data. A scatterplot or histogram that shows you that there is a clear pattern in the data. Maybe a simple model based on crude preliminary features that nonetheless works. Sometimes an analysis is doomed to failure, because there just isn't much signal in the data. If this is the case, you want to know sooner rather than later, so that you can change your focus.
- Automate your analysis in a single script so that it's easy to run the whole thing with one command. This is a point that I've learned the hard way: it is really, really easy after several hours at the command line to lose track of exactly what processing you did to get your data into its current form. Keep things reproducible from the beginning.
- Keep your code modular and broken out into clear stages. This makes it easy to modify, add in, and take out steps as you experiment.

Notice how much of this comes down to considerations of software, not analytics. The code must be flexible enough to solve all manner of problems, powerful enough to do it efficiently, and comprehensible enough to edit quickly if objectives change. Doing this requires that data scientists use flexible, powerful programming languages, which I will discuss in the next chapter.

2.10 Glossary

Data wrangling The nitty-gritty task of cleaning data and getting it into a standard format that is suitable for downstream analysis.

Exploratory analysis A stage of analysis that focuses on exploring the data to generate hypotheses about it. Exploratory analysis relies heavily on visualizations.

Feature A small piece of data, usually a number or a label, that is extracted from your data and characterizes some entity in your dataset.

Product requirements document (PRD) A document that specifies exactly what functionality a planned product should have.

Production code Software that is run repeatedly and maintained. It especially refers to source code of software product that is distributed to other people.

Statement of work (SOW) A document that specifies what work is to be done in a project, relevant timelines, and specific deliverables.

Target variable A feature that you are trying to predict in machine learning. Sometimes, it is already in your data, and other times, you must construct it yourself.

3

Programming Languages

One of the most obvious things that separate data scientists from traditional business analysts and (to a lesser degree) statisticians is that they spend a lot of their time writing code in a more-or-less normal programming language, as software engineers do. Sometimes, it's a statistically oriented language such as R, but even that is a far cry from something such as Excel or a graphical package such as Tableau.

This chapter will discuss why that is and give a brief survey of some of the more popular languages. It will then dive into the weeds of Python, my personal language of choice and the most popular option among data scientists. If you already know Python and its technical libraries, then feel free to skim. If not though, then this chapter will give you the foundation in Python to understand the example code in the rest of the book.

3.1   Why Use a Programming Language? What Are the Other Options?

To date, I have never worked on a data science project that could be done completely within a graphical package such as Excel or Tableau. There is always something – a weird formatting issue that requires coding up the edge cases, a dataset that's too large to fit into memory, an unconventional feature that I want to extract, or something else – that forces me to roll up my sleeves and write some code.

This will be your experience too, almost certainly. To put it glibly, data science is Turing complete. Many data scientists (like me) find it's more expedient to just work completely in programming languages, supplemented by numerical libraries. Others though find that it's worthwhile to do their data wrangling and feature extraction in a programming language but then load the datasets into another tool for their exploratory analysis.

The Data Science Handbook, First Edition. Field Cady.
© 2017 John Wiley & Sons, Inc. Published 2017 by John Wiley & Sons, Inc.

Here are some tools besides programming languages that you might want to incorporate into your workflow:

- Excel. Microsoft products often get a bad rap in the data science world, and it is completely undeserved. For simple data analysis, Excel is probably the best piece of software ever made.
- Tableau. This is a tool for visualizing the data in relational databases. It's pretty limited in its functionality in my experience, but when it works, the graphics are absolutely beautiful.
- Weka. This is a tool for applying pre-canned machine learning algorithms to datasets that are already well formatted and contain the relevant features. An advantage of Weka is that it's really just a thin GUI wrapper around some Java libraries, so it's easy to use the same models in your exploratory analysis and later production code (assuming that you work in Java).

There is something critical that all of these tools have in common: they assume that your data is already in tabular form! Tabular datasets, and the things you can do with them, are sufficiently standardized that people can easily write reusable tools that streamline the most common operations.

However, each dataset often requires its own idiosyncratic data wrangling. Furthermore, each new problem will require creativity and flexibility in what features you extract from the raw data – especially if your raw data is very far from tabular form. This is the reason that every data scientist needs to be proficient in at least one programming language.

3.2 A Survey of Programming Languages for Data Science

There are many programming language options available for data scientists. This section will give you a run-down of some of the most popular ones.

3.2.1 Python

The example code in this book is generally in Python, for a number of reasons. In my opinion it is the best programming language available for general-purpose use, but that's largely a matter of personal taste. It is also a very popular choice among data scientists, who feel like it balances the flexibility of a conventional scripting language with the numerical muscles of a good mathematics package (at least, when it's paired with its scientific computing libraries).

Python was developed by Guido van Rossum and first released in 1991. The language itself is a high-level scripting language, with functionality similar to Perl and Ruby and with an unusually clean and self-consistent syntax. Outside of the core language, Python has several open-source technical computing libraries that make it a powerful tool for analytics.

3.2.2 R

Aside from Python, R is probably the most popular programming language among data scientists. Python is a scripting language designed for computer programmers, which has been augmented with libraries for technical computing. In contrast, R was designed by and for statisticians, and it is natively integrated with graphics capabilities and extensive statistical functions. It is based on S, which was developed at Bell Labs in 1976.

R was brilliant for its time and a huge step up from the Fortran routines that it was competing with. In fact, many of Python's technical computing libraries are just ripping off the good ideas in R. But almost 40 years later, R is showing its age. Specifically, there are areas where the syntax is very clunky, the support for strings is terrible, and the type system is antiquated.

In my mind, the main reason to use R is just that there are so many special libraries that have been written for it over the years, and Python has not covered all the little special use cases yet. I no longer use R for my own work, but it is a major force in the data science community and will continue to be for the foreseeable future. In the statistics community, R is still the lingua franca. You should know about it, even if you don't use it yourself.

3.2.3 MATLAB® and Octave

The data science community skews strongly toward open-source software, so good proprietary programs such as MATLAB® often get less credit than they deserve. Developed and sold by the MathWorks Corporation, MATLAB® is an excellent package for numerical computing. It has a more consistent (and, in my opinion, nicer) syntax compared to R and more numerical muscle compared to Python. A lot of people coming from physics or mechanical/electrical engineering backgrounds are well-versed in MATLAB®. It is not as well-suited to large software frameworks or string-based data munging, but it is best-in-class for numerical computing.

If you like MATLAB's syntax, but don't like paying for software, then you could also consider Octave. It is an open-source version of MATLAB®. It doesn't capture all of MATLAB's functionality and certainly doesn't have the same support infrastructure, but it's a fine option.

3.2.4 SAS®

SAS (Statistical Analysis Software) is a proprietary statistics framework that dates back to the 1960s. Similar to R, there is a tremendous amount of entrenched legacy code written in SAS and a wide range of functionality that has been put into it. However, the language itself is very alien to somebody more used to modern language. SAS can be great for the business statistics applications that it is so popular in, but I don't recommend it for general-purpose data science.

3.2.5 Scala®

Scala is an up-and-coming language that shows a lot of promise. It is not currently a suitable general-purpose tool for data scientists because it doesn't have the library support for analytics and visualizations. However, that could easily change in the same way that it did with Python. Scala is similar to Java under the hood, but has a much simpler syntax with a lot of powerful features borrowed from other languages (especially functional languages). It works both for general-purpose scripting and for large-scale production software. Many of the most popular Big Data technologies are written in Scala.

3.3 Python Crash Course

This section will give a quick tutorial on the Python language. My goal is to get you up-and-running quickly with the basics of the language, especially so that you can understand the example code in the book.

The tutorial is far from exhaustive. There are many aspects of Python that I don't discuss, and in particular, I ignore most of its many built-in libraries. Some of this material will be covered later in the book when it becomes relevant.

The next section will give you an introduction to Python's technical libraries, which elevate it from a solid scripting language to a one-stop-shop for data science.

3.3.1 A Note on Versions

There are a number of versions of the Python language out there. As of this writing, the Python 2.7 series was by far the most popular for data scientists. The main reason for this is that all of the numerical computing libraries work with it.

In 2008, Python 3.0 was released, and it broke backward compatibility with Python 2.7. This was a big deal because the Python community tends to be very careful about keeping things mutually consistent.

This book is written assuming Python 2.7, but most of what I will say applied equally well to 3.x. Several of the key places where 3.x differs are as follows:

- Print is treated as a function. So you would say

```
>>> print("hello world")
```

instead of

```
>>> print "hello world"
```

- Arithmetic operations are treated as decimal operations even when they are done on integers. That way

3 / 2

will equal 1.5 rather than 1. If you want to do integer division, then say //
- String and Unicode are removed as separate classes. It's all Unicode now, and you can use the ByteArray type if you want to manipulate bytes directly.

3.3.2 "Hello World" Script

A common way to learn a new programming language is to first write a "hello world!" program: this is a program that just prints the text "hello world!" to the screen. If you can write it and run it, then you know you have your software environment set up correctly and you know how to use it. After that point, you're ready to roll with the serious code.

There are two ways you can run Python code, and I'll walk you through hello world in both of them. Either you can open up the Python interpreter and enter your commands one at a time, which is very useful for exploring data and experimenting with what you want to do, or you can put your code into a file and run it all at once.

To run code in the interpreter on a Mac or Linux system, do the following:

1) Go to the command terminal.
2) Type "python" and press enter. This will display the command prompt >>>.
3) Type "print 'hello world'" and press enter. The phrase "hello world" should print on the screen.
4) The whole thing should appear as follows:

```
>>> print "hello world!"
hello world!
```

5) Congratulations! You've just run a line of Python code.
6) Press Ctrl-d to close the interpreter.

The process is very similar if you are working in a Windows environment. In place of the command terminal you are likely to use PowerShell — it is the Windows equivalent of a bash terminal. For editing your source code, Visual Studio is a powerful IDE that is ubiquitous among Windows programmers. Personally I tend to write my scripts in plain text editors if it's at all practical, but especially for larger codebases a good IDE becomes invaluable.

3.3.3 More Complicated Script

Ok, now that you've got Python running, let's jump into the deep end. Here is a more complicated Python script. It has a data structure that describes employees of a company. It goes through the employee records, gives each one a 5% raise, and updates the record with the name of the state they live in. It then prints out information describing the updated employee data. Don't worry if you can't read the whole thing right now: I'll explain what all the parts are. After we walk through this script, I'll give a more comprehensive overview of Python's data types and how to work with them; the script doesn't show it all.

```python
SALARY_RAISE_FACTOR = 0.05
STATE_CODE_MAP = {'WA': 'Washington', 'TX': 'Texas'}

def update_employee_record(rec):
    old_sal = rec['salary']
    new_sal = old_sal * (1 + SALARY_RAISE_FACTOR)
    rec['salary'] = new_sal
    state_code = rec['state_code']
    rec['state_name'] = STATE_CODE_MAP[state_code]

input_data = [
    {'employee_name': 'Susan', 'salary': 100000.0,
     'state_code': 'WA'},
    {'employee_name': 'Ellen', 'salary': 75000.0,
     'state_code': 'TX'},
]

for rec in input_data:
    update_employee_record(rec)
    name = rec['employee_name']
    salary = rec['salary']
    state = rec['state_name']
    print name + ' now lives in ' + state
    print '   and makes $' + str(salary)
```

If you run this script, you will see the following output:

```
Susan now lives in Washington
   and makes $110250.0
Ellen now lives in Texas
   and makes $78750.0
```

The first line of the script defines the variable SALARY_RAISE_FACTOR to be the decimal number 0.05.

The next line defines what's called a dict (short for dictionary) called STATE_CODE_MAP, which maps the postal abbreviations of several states to their full names. A dict maps "keys" to "values," and they are enclosed within curly braces. There are commas between each key/value pair, and the key and value are separated by a colon. The keys in a dict are usually strings, but they can also be numbers or any other atomic-type except None (which we'll see in a minute). The values can be any Python object whatsoever, and different values can have different types. But in this case, the values are all strings. Dicts are one of Python's three main "container" data types (i.e., it contains other data), the other two being lists and tuples.

Next up, the line

```
def update_employee_record(rec):
```

says that we are defining a function called `update_employee_record` that takes in a single argument and that the argument is called rec within the scope of this function. In our code, `rec` will always be a dict, but we have *not* specified that in the function declaration. You can pass an integer, string, or anything else into `update_employee_record`. In this case, it so happens that we later do operations to rec that will fail if it's not a dictionary (or something that behaves like one), but Python won't know anything is amiss until the operation fails.

Here we come to the most famous gotcha of Python. The rest of the body of the function is all indented the same way: exactly four spaces. It could have been two spaces, or a tab, or any other whitespace combinations, but it *must* be consistent. Consistency such as this is good practice in any programming language since it makes the code easier to read, but Python requires it. This is the single most controversial thing about Python, and it can get confusing if you're in a situation where tabs and spaces are getting mixed.

In the body of the function, when we say

```
old_sal = rec['salary']
```

we are pulling out the "salary" field in rec. This is how you get data out of a dict. By doing this, we are also tacitly assuming that there *is* a "salary" field: the code will throw an error if there isn't one. Later when we say

```
rec['salary'] = new_sal
```

we are assigning to the "salary" field – creating the field if there isn't one, and overwriting it if there's already one there.

The `input_data` variable is a list. Lists can contain elements of any type, but in this case, they are all dictionaries. Note that in this case, the values in the dictionaries are not all the same type: some are strings, but there is also a float field.

In the last part of the script, the line

```
for rec in input_data:
```

will loop over all of the elements of `input_data` in a order, executing the body of the loop for each one. Similar to function declarations, the body of the loop *must* be indented consistently.

The print statement here deserves a special mention. When we say

```
print '   and makes $' + str(salary)
```

there are three things going on:

- str(salary) takes salary, which is a float like 75,000.0, and returns a string like "75,000". str() is a function that takes many Python objects and returns a string representation of them.
- Adding two strings with + just concatenates them. Adding a string to a float would have given an error, which is why we had to say str(salary).
- The print statement in Python is a little weird. Most built-in functions in Python are called with parentheses, but print doesn't use them. This rare inconsistency was remedied in Python 3.0.

3.3.4 Atomic Data Types

Python has five main atomic data types that you'll have to worry about. If you have used a programming language in the past, they should mostly sound pretty familiar:

- int: A mathematical integer
- float: A floating-point number
- bool: A true/false flag
- string: A piece of text of arbitrarily many characters (possibly 0 or 1)
- NoneType: This is a special type with only a single value None. It is often used as a placeholder when there is missing data or some process failed.

Declaring a variable that is an int or a float is very straightforward:

```
my_integer = 2
my_other_integer = 2 + 3
my_float = 2.0
```

Boolean values are similarly uncomplicated:

```
my_true_bool = True
my_false_bool = False
this_is_true = (0 < 100)
this_is_false = (0 > 100)
```

NoneType is special. The only value it can take is called None, and this is often used as a placeholder when a variable should exist, but you don't want it to have a meaningful value. Functions that fail in some way will also often return None to signify that there was a problem.

3.4 Strings

By far the most complicated of the atomic data types is string. A string is a piece of text of arbitrary length.

You declare a string by enclosing the text in quotation marks. You can use single quotes or double quotes – they're equivalent to each other. However, you might want to enclose your string in double quotes if the string contains the single quote character, and vice versa.

```
a_string = "hello"
same_as_previous = 'hello'
an_empty_string = ""
w_a_single_quote = "hello's"
```

In place of a single quotation character, you can also enclose a string in a triple of characters. Unlike normal strings, one enclosed in triple quotes is also allowed to extend over multiple lines

```
multi_line_string = """line 1
line 2"""
```

It's common to use triple quoted strings for things such as large pieces of text that are embedded in your code (say, some HTML that you're using a lot because your script is writing an HTML document).

If you want to put special characters, such as a tab, a newline, or a weird hex code into your string, you can do it by a process called "escaping." When the "\" character is written in the string, it and the next character together encode a single nonstandard character. The most common of these are the new line "\n", the tab "\t", and the slash character itself "\\."

To take a substring of a string in Python, you use bracket notation as follows:

```
>>> "ABCD"[0]
'A'
>>> "ABCD"[0:2]
'AB'
>>> "ABCD"[1:3]
'BC'
```

If you want to pull a single character out of a string, then you can do it with brackets such as this, passing in the index of the character you want:

```
>>> "ABCD"[0]
'A'
```

Note that the indices start at 0, not 1. If you want a substring of length greater than 1, you put starting and ending indices in the brackets:

```
>>> "ABCD"[0:2]
'AB'
>>> "ABCD"[1:3]
'BC'
```

The first number says which index to start at, and the second number says which index to stop just short of. If the first number is omitted, then you will start at the beginning. If the second is omitted, you will continue to the end. So we can say

```
>>> "ABCD"[1:]
'BCD'
```

You can also use negative indices. −1 will refer to the last element in the list, −2 to the one before, etc. So you can drop the last character in a string such as this:

```
>>> "ABCD"[:-1]
'ABC'
```

The next chapter will go into a lot more detail about the various tools that Python has for working with strings.

3.4.1 Comments and Docstrings

There are two kinds of comments in Python:

- Those denoted by a # character, such as this:

  ```
  # This whole line is a comment
  a = 5  # and the last part of this line is too
  ```

- Strings that take up a line (or more) in your code but aren't assigned to a variable.

It's common practice to have a string at the beginning of a Python file that describes what the file does and how to use it. Such a string is called a docstring. If you import the file as a library, then that library will have a field called __doc__ that acts as built-in documentation. These things come in handy! A function can also have a doc string, such as this:

```
def sqr(x):
    "This function just squares its input "
    return x * x
```

3.4.2 Complex Data Types

Python has three main data containers: lists, tuples, and dicts. There is also one called a set that you will use less often. Each of them contains other data structures, hence the name.

The first thing to know about containers in Python is that, unlike many other languages, you can mix and match the types. A list can consist entirely of ints. But it can also contain tuples, dictionaries, user-defined types, and even other lists.

All of Python's container types are classes, in the object-oriented sense. However, they also all act as functions, which try to coerce their arguments into the appropriate type. For example,

```
my_list = ["a ", "b ", "c "]
my_set = set(my_list)
my_tuple = tuple(my_list)
```

will create a list and then create a set and a tuple that contain identical data.

3.4.3 Lists

A list is just what it sounds like: an ordered list of variables. The following code shows basic usage:

```
my_list = ["a ", "b ", "c "]
print my_list[0] # prints "a "
my_list[0] = "A " # changes that element of the list
my_list.append("d ") # adds new element to the end
# List elements can be ANYTHING
mixed_list = ["A ", 5.7, "B ", [1,2,3]]
```

There is a special operation called a list comprehension, which lets us create one list from another by applying the same operation to all of its elements (and possibly filtering out some of those elements):

```
original_list = [1,2,3,4,5,6,7,8]
squares = [x*x for x in original_list]
squares_of_evens = [x*x for x in original_list
  if x%2==0]
```

If you have not seen it before, there is one very important convention with list indexing that can be confusing at first: the first element in the list is element number 0. The second is element number 1, and so on. There are reasons (some of them historical) for this convention, and if it mystifies you, you're not alone. But you will have to get used to it with Python.

If you want to select a subset of a list, then you can do it with a colon:

```
my_list = ["a", "b", "c"]
first_two_elements = my_list[0:3]
```

The first number says which index to start at, and the second number says which index to stop just short of. If the first number is omitted, then you will start at the beginning. If the second is omitted, you will continue to the end. So we can say

```
my_list = ["a", "b", "c"]
first_two_elements = my_list[:3]
last_two_elements = my_list[1:]
```

You can also use negative indices. −1 will refer to the last element in the list, −2 to the one before, etc. So we can say

```
my_list = ["a", "b", "c"]
all_but_last_element = my_list[:-1]
```

3.4.4 Strings and Lists

For complex string manipulation, one of the most flexible methods that you can call on strings is split(). It will break a string up on whitespace and return those parts of it as a list. Alternatively, you can pass another string as an argument, which will cause you to split on that string instead. It works such as this:

```
>>> "ABC DEF".split()
['ABC', 'DEF']
>>> "ABC \tDEF".split()
['ABC', 'DEF']
>>> "ABC \tDEF".split(' ')
['ABC', '\tDEF']
>>> "ABCABD".split("AB")
['', 'C', 'D']
```

The inverse of split() is the join() method. It is called on a string, and you pass in a list of other strings. The strings are all then concatenated into one string, using the string as a delimiter. For example,

```
>>> ",".join(["A", "B", "C"])
'A,B,C'
```

You might have noticed that the syntax for selecting characters in a string is the same as that for selecting elements in a list. In general, it is called "slice notation," and it is possible to create other Python objects that use the same notation. Most generally, a slice takes in a start index, an end index, and how big the spacing should be. For example,

```
>>> start, end, count_by = 1, 7, 2
>>> "ABCDEFG"[start: end: count_by]
'BDF'
```

3.4.5 Tuples

A tuple is conceptually a list that cannot be modified (no changing the elements, no adding/removing elements). Having them may seem redundant, but tuples are much more efficient than lists in some cases, and they play a central role in the operation of Python under the hood. There are also several things that, for technical reasons, you can do with tuples that you can't with lists. The most obvious of these is that the keys in a dictionary cannot be lists, but they can be tuples.

```
my_tuple = (1, 2, "hello world")
print my_tuple[0] # prints 1
my_tuple[1] = 5 # This will give an error!
```

There is one important piece of syntactic sugar to know that is often used with tuples. Oftentimes, we want to give names to the different fields in a tuple, and it is clunky to explicitly define a new variable for each of them. In these cases, we can do multiple assignment as follows:

```
my_tuple = (1, 2)
zeroth_field, first_field = my_tuple
```

3.4.6 Dictionaries

A dictionary is a structure that takes in a key and returns a value. The keys for a dictionary are usually strings, but they can also be any other atomic data type or tuples (but they can't be lists or dictionaries). The values can be anything at all – integers, other dictionaries, external libraries, etc. In defining a dictionary, you use curly braces, with a colon separating the key and its value:

```
my_dict = {"January": 1, "February":2}
print my_dict["January"] # prints 1
my_dict["March"] = 3 # add new element
my_dict["January"] = "Start of the year" # overwrite
old value
```

As an interesting note, the Python language itself is largely built out of dictionaries (or slight variations of them). The namespace that stores all of your variables, for example, is a dictionary mapping the variables' names to the objects themselves.

You can also create a dictionary by passing in a list of tuples to the function dict(), and you can create a list of tuples by calling the items() method on a dictionary:

```
pairs = [("one",1), ("two",2)]
as_dict = dict(pairs)
same_as_pairs = as_dict.items()
```

3.4.7 Sets

A set is somewhat similar to a dictionary with only keys and no values. It stores a collection of unique objects that are of atomic types. You can add new values to a set, which will do nothing if the value is already in it. You can also query the set to see if a value is in it. A simple shell script shows how this works:

```
>>> s = set()
>>> 5 in s
False
>>> s.add(5)
>>> 5 in s
True
>>> s.add(5) # does nothing
```

3.5   Defining Functions

A function in Python is defined and called as follows:

```
def my_function(x):
 y = x+1
 x_sqrd = x*x
 return x_sqrd

five_plus_one_sqrd = my_function(5)
```

This is a so-called "pure function," meaning that it takes some input, returns an output, and does nothing else. A function can also have side effects, such as printing something to the screen or operating on a file. In our example script earlier, modifying the input dictionary was a side effect. If no return value is specified, the function will return None.

You can also define optional arguments in a function, using this syntax:

```
def raise(x, n=2):
 return pow(x,n)
```

```
two_sqrd = raise(2)
two_cubed = raise(2, n=3)
```

If the function you are defining only contains one line, and has no side effects, you can also define it using a so-called lambda expression:

```
sqr = lambda x : x*x
five_sqrd = sqr(5)
```

Assigning a lambda expression to "sqr" is equivalent to the normal syntax for function definitions. The term "lambda" is a reference to the Lisp programming language, which defines functions using the "lambda" keyword in a similar way.

Lambda functions are mostly used if you're passing a one-off function as an argument to another function, and there's no need to pollute the namespace with a new function name. For example,

```
def apply_to_evens(a_list, a_func):
 return [a_func(x) for x in a_list if x%2==0]
my_list = [1,2,3,4,5]
sqrs_of_evens = apply_to_evens(my_list, lambda x:x*x)
```

Functions such as this, which are defined on the fly and never given an actual name, are called "anonymous functions." They can be very handy in data science, especially in Big Data.

3.5.1 For Loops and Control Structures

The main control structure you do in practice is to loop over a list, as follows:

```
my_list = [1, 2, 3]
for x in my_list:
 print "the number is ", x
```

If you are iterating over a list of tuples (as you might if you're working with a dictionary), you can use the shorthand tuple notation I mentioned previously:

```
for key, value in my_dict.items():
 print "the value for ", key, " is ", value
```

More generally, any data structure that allows for-loops such as this is called "iterable." Lists are the most prominent iterable data type, but they are far from the only one.

If statements are handled this way,

```
if i < 3:
 print "i is less than three"
elif i < 5: print "i is between 3 and 5"
else: print "i is greater than 5"
```

You don't see it that often in practice, but Python also allows for while-loops, which are similar to this:

```
i = 0
while i < 5:
 print "i is still less than five"
 i = i+1
```

3.5.2 A Few Key Functions

Python has a small number of built-in functions that you should be aware of.

Function name	Action	Examples

int	Cast to an int	int(5.7) # rounds down
		int("5")

float	Cast to a float	float(5) float("5.7")
bool	Cast to a bool	bool("") # False bool("asdf") # True
str	Cast to a str	
dict	Turns list of key/value tuples into dictionary	dict([("January", 1), ("February", 2)])
range	Range(n) gives a list of integers from 0 to $n-1$. That is, it starts at 0 and has length n	range(5) # 0 to 4 range(4,18) # 4 to 17
zip	Take in two lists and pair off the elements into one list of tuples	zip(["Sunday", "Monday", "Tuesday"], range(3))

open	Opens a text file for reading or writing. The second argument is an "r" for reading the file and a "w" for writing it. You can also use "a" to just append to the end of a file	# get file contents
		# as one big string
		open("file.txt", "r").read()
		# Get file contents
		# as list of strings
		open("file.txt", "r").readlines()
		# Write
		open("file.txt", "w").write("Hello world! ")

Function name	Action	Examples
len	Give the length of something. For a list or tuple, it will be the length. For a string, it will be the number of characters	len("sdf") # 3 len([1,2,3,4]) # 4

enumerate	Pass in some indexable object (usually a list). Get out index/value tuples, which give indices in the object and their corresponding values. Useful if you are looping over a list, but you also need to keep track of the index	`for ind, val in mylist:`
		`    print "At %i" % i`
		`    print val`

3.5.3 Exception Handling

If Python code fails, sometimes, we want to have the script be prepared for that and act accordingly (rather than just dying). That is illustrated here:

```
try:
    lines = input_text.split("\n")
    print "tenth line was: ", lines[9]
except:
    print "There were < 10 lines"
```

3.5.4 Libraries

To import functionality from an existing library, you use any of the following syntax:

```
from my_lib import f1, f2 # f1 & f2 in namespace
import other_lib as ol # ol.f1 is the f1 func
from other_lib import * # f1 is in namespace
```

Generally, the first and second methods of importing a library make for the most readable code; if you import * from several libraries and then call f1 later on in your code, it's not obvious which library f1 came from.

To write your own library, just write a .py file in which your functions, classes, or other objects are defined. It can then be imported using the aforementioned syntax. Just make sure that your library is in the directory you are running your code from or in some other place that Python can find it.

3.5.5 Classes and Objects

Strictly speaking, everything in Python (and I mean everything – integers, functions, classes, imported libraries, etc.) is what's called an object. However, most of the language is built around a few high-powered classes (such as lists and dictionaries) that do most of the heavy lifting, so it's common to use Python only as a scripting language.

However, if you want to define your own classes, you can do it this way:

```
class Dog:
 def __init__(self, name):
   self.name = name
 def respond_to_command(self, command):
   if command == self.name: self.speak()
 def speak(self):
   print "bark bark!!"

fido = Dog("fido")
fido.respond_to_command("spot") # does nothing
fido.respond_to_command("fido") # prints bark bark
```

Here __init__ is a special function that gets called whenever an instance of the class is created. It does all of the initial setup required for the object.

The one thing that throws a lot of people off is the "self" keyword that gets passed in as the first argument for every function in the class. When I call fido. respond_to_command, the "self" argument in respond_to_command refers to fido himself, that is, the Dog object whose method is being called. This allows us to refer specifically to fido's data elements, such as self.name. For many object-oriented languages, just saying "name" in resond_to_command will implicitly refer to fido's name, but Python requires that it be explicit. It's similar to the keyword "this" that you will see in languages such as C++.

3.5.6 GOTCHA: Hashable and Unhashable Types

When I first started learning Python, there was one big gotcha that I ran into. It caused me a lot of grief for a few days as I tried to figure out why my code was failing, and I would like to spare you my pain. Python's data type falls into two categories:

- Hashable types. This includes ints, floats, strings, tuples, and a few more obscure ones. These are generally low-level data types, and instances of them are immutable.
- Unhashable types include lists, dictionaries, and libraries. Generally, unhashable types are for larger, more complex objects, which have internal structure that can be modified.

The biggest difference between hashable and unhashable types is illustrated in this shell session:

```
>>> a = 5   # a is a hashable int
>>> b = a   # b points to a COPY of a
```

```
>>> a = a + 1
>>> print b    # b has NOT been incremented
5
>>> A = []     # A is an UNhashable list
>>> B = A      # B points to the SAME list as A.
>>> A.append(5)
>>> B
[5]
```

When I say b = a, a copy of the hashable int is made in memory, and the variable name b is set to point to it. But when I'm using unhashable lists and say B = A, the variable B is set to point to the exact same list!

If I had truly wanted to make a copy of A, so that appending to A didn't affect B, I could have said something like the following:

```
>>> B = [x for x in A]
```

which would have constructed a new list in memory. If A was a list of integers, then A and B would be incapable of stepping on each other's toes: they would have their own separate copies of the numbers.

However, if the elements of A were themselves unhashable types, then B would be distinct from A, but they would be pointing to the same objects. For example,

```
>>> A = [{}, {}]   # list of dicts
>>> B = [x for x in A]
>>> A[0]["name"] = "bob"
>>> B[0]["name"]
"bob"
```

The other thing about hashable types is that the keys in a dictionary must be hashable.

3.6   Python's Technical Libraries

Python was designed mostly as a tool for software engineers, but there is an excellent suite of libraries available that make it a first-class environment for technical computing, competing with the likes of MATLAB® and R. The main ones, which will be covered in this book, are as follows:

- Pandas: This is the big one for you to know. It stores and operates on data in data frames, very efficiently and with a sleek, intuitive API.
- NumPy: This is a library for dealing with numerical arrays in ways that are fast and memory efficient, but it's clunky and low level for a user. Under the hood, Pandas operates on NumPy arrays.

- Scikit-learn: This is the main machine learning library, and it operates on NumPy arrays. You can take Pandas objects, turn them into NumPy arrays, and then plug them into scikit-learn.
- Matplotlib: This is the big plotting and visualization library. Similar to NumPy, it is low level and a bit clunky to use directly. Pandas provides human-friendly wrappers that call matplotlib routines.
- SciPy: This provides a suite of functions that perform fancy numerical operations on NumPy arrays.

These aren't the only technical computing libraries available in Python, but they're by far the most popular, and together they form a cohesive, powerful tool suite.

NumPy is the most fundamental library; it defines the core numerical arrays that everything else operates on. However, most of your actual code (especially data munging and feature extraction) will be working within Pandas, only switching to the other libraries as needed. The rest of this chapter will be a quick crash course on the basic data structures of Pandas.

3.6.1 Data Frames

The central kind of object in Pandas is called a DataFrame, which is similar to SQL tables or R data frames. A data frame is a table with rows and columns, where each column holds data of a particular type (such as integers, strings, or floats). DataFrames make it easy and efficient to apply a function to every element in a column or to calculate aggregates such as the sum of a column. Some of the basic operations on data frames are shown in this code:

```
import pandas as pd
# Making data frame from a dictionary
# that maps column names to their values
df = pd.DataFrame({
  "name": ["Bob", "Alex", "Janice"],
  "age": [60, 25, 33]
  })
# Reading a DataFrame from a file
other_df = pd.read_csv("myfile.csv")

# Making new columns from old ones
# is really easy
df["age_plus_one"] = df["age"] + 1
df["age_times_two"] = 2 * df["age"]
df["age_squared"] = df["age"] * df["age"]
df["over_30"] = (df["age"] > 30) # this col is bools

# The columns have various built-in aggregate functions
```

```
total_age = df["age"].sum()
median_age = df["age"].quantile(0.5)

# You can select several rows of the DataFrame
# and make a new DataFrame out of them
df_below50 = df[df["age"] < 50]
# Apply a custom function to a column
df["age_squared"] = df["age"].apply(lambda x: x*x)
```

One important thing about DataFrames is the notion of an index. This is basically a name (not necessarily unique) that is given to every row of the data frame. By default, the indexes are just the line numbers (starting at 0), but you can set the index to be other columns if you like:

```
df = pd.DataFrame({
  "name": ["Bob", "Alex", "Jane"],
  "age": [60, 25, 33]
  })
print df.index # prints 0-2, the line numbers

# Create a DataFrame containing the same data,
# but where name is the index
df_w_name_as_ind = df.set_index("name")
print df_w_name_as_ind.index # prints their names

# Get the row for Bob
bobs_row = df_w_name_as_ind.ix["Bob"]
print bobs_row["age"] # prints 60
```

3.6.2 Series

Besides DataFrames, the other big data structure in Pandas is the Series. Really, I've already shown them to you: a column in a DataFrame is a Series. Conceptually, a Series is just an array of data objects, all the same type, with an index associated. The columns of a DataFrame are Series objects that all happen to share the same index.

The following code shows you some of the basic Series operations, independent of their function in DataFrames:

```
>>> # import Pandas.  I always alias it as pd
>>> import pandas as pd
>>> s = pd.Series([1,2,3])  # make Series from list
>>>
>>> # display the values in s
>>> # note index is to the far left
>>> s
```

```
0    1
1    2
2    3
dtype: int64
>>> s+2   # Add a number to each element of s
0    3
1    4
2    5
dtype: int64
>>> s.index   # you can access the index directly
Int64Index([0, 1, 2], dtype='int64')
>>> # Adding two series will add corresponding
elements to each other
>>> s + pd.Series([4,4,5])
0    5
1    6
2    8
dtype: int64
```

Now technically, I lied to you a minute ago when I said that a Series object's elements all have to be the same type. They have to be the same type if you want all the performance benefits of Pandas, but we have actually already seen a Series object that mixes its types:

```
>>> bobs_row = df_w_name_as_ind.ix["Bob"]
>>> type(bobs_row)
<class 'pandas.core.series.Series'>
>>> bobs_row
age                   60
age_plus_one          61
age_times_two        120
age_squared         3600
over_30             True
Name: Tom, dtype: object
```

So we can see that this row of a data frame was actually a Series object. But instead of int64 or something similar, its dtype is "object." This means that under the hood, it's not storing a low-level integer representation or anything similar; it's storing a reference to an arbitrary Python object.

3.6.3 Joining and Grouping

So far we've focused on the following DataFrame operations:

- Creating data frames

- Adding new columns that are derived from basic operations on existing columns
- Using simple conditions to select rows in a DataFrame
- Aggregating columns
- Setting columns to function as an index, and using the index to pull out rows of the data.

This section discusses two more advanced operations: joining and grouping. These may be familiar to you from working with SQL.

Joining is used if you want to combine two separate data frames into a single frame containing all the data. We take two data frames, match up rows that have a common index, and combine them into a single frame. This shell session shows it:

```
>>> df_w_age = pd.DataFrame({
  "name": ["Tom", "Tyrell", "Claire"],
  "age": [60, 25, 33]
  })
>>> df_w_height = pd.DataFrame({
  "name": ["Tom", "Tyrell", "Claire"],
  "height": [6.2, 4.0, 5.5]
  })
>>> joined = df_w_age.set_index("name").join(
            df_w_height.set_index("name"))
>>> print joined
         age   height
name
Tom       60     6.2
Tyrell    25     4.0
Claire    33     5.5
>>> print joined.reset_index()
     name   age   height
0    Tom     60     6.2
1    Tyrell  25     4.0
2    Claire  33     5.5
```

The other thing we often want to do is to group the rows based on some property and aggregate each group separately. This is done with the groupby() function, the use of which is shown here:

```
>>> df = pd.DataFrame({
  "name": ["Tom", "Tyrell", "Claire"],
  "age": [60, 25, 33],
    "height": [6.2, 4.0, 5.5],
```

```
      "gender": ["M", "M", "F"]
  })
>>> # use built-in aggregates
>>> print df.groupby("gender").mean()
        age   height
gender
F       33.0     5.5
M       42.5     5.1
>>> medians = df.groupby("gender").quantile(0.5)
>>> # Use a custom aggregation function
>>> def agg(ddf):
        return pd.Series({
            "name": max(ddf["name"]),
            "oldest": max(ddf["age"]),
            "mean_height": ddf["height"].mean()
            })
>>> print df.groupby("gender").apply(agg)
        mean_height    name   oldest
gender
F              5.5   Claire      33
M              5.1      Tom      60
```

3.7 Other Python Resources

One of the benefits of using Python is that there is a huge amount of very clear documentation available online. It's extremely easy to just google around and find the right syntax or libraries to do whatever it is you need to get done. Besides just searching around, I recommend the following resources:

- https://docs.python.org/2/ This is the main resource for documentation of Python version 2's syntax.
- http://pandas.pydata.org/ The official documentation for the pandas library.
- http://scikit-learn.org/stable/index.html The documentation for scikit-learn. This is some of the best documentation I've ever seen for software. Most of it is example scripts that show off all the various things you can do with scikit-learn.

3.8 Further Reading

One of the benefits of learning Python (or any programming language) is that there is a huge amount of very clear documentation available online. It's extremely easy to just search around and find the right syntax or libraries to do whatever it is you need to get done.

Besides general browsing, I can recommend several specific resources that are great for coming up to speed:

1 Pilgrim, M, 2004, *Dive into Python: Python from Novice to Pro*, viewed 7 August 2016, http://www.diveintopython.net/.

2 *Pandas: Python Data Analysis Library*, viewed 7 August 2016, http://pandas.pydata.org/.

3 https://www.python.org/, viewed 7 August 2016, The Python Software Foundation.

4 Scott, M, *Programming Language Pragmatics*, 4th edn, 2015, Morgan Kaufmann, Burlington, MA.

3.9 Glossary

Anonymous function A function that is never given a name.

DataFrame The main Pandas data structure. It stores a dataset as a table with rows and columns.

Dict A Python object, which maps keys (which must be of a hashable type) to values (which can be any type).

Hashable type ints, floats, strings, and a couple other low-level Python data types.

Index Identifiers for each row in a DataFrame or element in a Series.

Join An operation that takes two data frames and concatenates matching rows into a large data frame. Rows match if they have the same entries in whatever column you are joining on.

List A Python object that stores an ordered list of objects. It is an unhashable type, so we can do things such as appending now elements.

NumPy A low-level Python library for efficiently processing numerical arrays.

Pandas A high-level Python library for manipulating data. It defines the DataFrame and Series types and is implemented using NumPy under the hood.

Pure function A function with no side effects.

Series A Pandas data type for storing a sequence of objects. The columns of a DataFrame are actually Series objects.

Set A Python container type that acts as a mathematical set.

Side effect A modification that is made to an existing object in memory, as opposed to creating a new object while leaving existing ones intact. Operations such as print to the screen and file interactions are also side effects.

Tuple A Python object that stores an ordered sequence of objects. Unlike lists, tuples are immutable and hashable.

Unhashable type Any Python type that is not hashable. Examples include lists, dicts, and user-defined classes. When you assign an unhashable object to a variable name, you will get a pointer to the original object, rather than to a copy of it.

Interlude: My Personal Toolkit

Every data scientist has their own set of preferred programming languages, libraries, and other tools. You will have to decide what works best for you. To give you a data point though, here is how I work when do data analysis:

- My main programming language for data science is Python. I know it, I love it, and I can do just about anything with it. I also use it for production coding whenever I am choosing the tools and there's no good reason not to.
- I use Pandas as my main data analysis library, and I supplement it with scikit-learn for machine learning.
- I usually use matplotlib for visualizations, but I'm looking to branch out. In particular, bokeh is an extremely promising recent arrival to the visualization scene. It is designed particularly for making interactive graphs that you access with a web browser.
- A lot of people use an Integrated Development Environment (IDE) for Python, such as Spyder or PyCharm. Personally though, I'm a little old school: I open up Python from the command line, and I edit my scripts in a plain text editor such as Sublime or TextWrangler. I'm considering switching to a browser-based notebook though, such as Jupyter.
- I do most of my work on a Mac, but that's just because it's what my employers tend to use. I usually do hobby projects using Linux, and I'm hoping to do more work on Windows in the future because they have a famously great set of tools for developers.
- When I'm doing Big Data I use PySpark, which I'll talk about in the chapter on Big Data.

I used to use R, but not anymore if I can avoid it. The syntax has always annoyed me, but the breaking point came a few years ago. I had an R script operating on a massive dataset that I had run and debugged several times, and it would always fail several hours in because of some memory issues. It was extremely frustrating, and I was getting close to a deadline. So, finally, I gave up on R and rewrote the entire thing in Python; it finished in 45 minutes the first time I ran it. In fairness, I wrote my Python code to be pretty efficient in its

The Data Science Handbook, First Edition. Field Cady.

memory usage, and my R code had used the notoriously inefficient plyr library, but it left a lasting impression.

By the time you read this, my toolkit may have changed. New toys are constantly becoming available, and it's important to stay abreast of them. Some people are forever trying out the newest libraries, always eager to find slightly better ways of doing things. Personally, I'm more inclined to wait until it's clear that a new tool is better before jumping on the bandwagon, so as not to spend a lot of time learning things that become obsolete. But no matter how you do it, one of the coolest parts of data science is the constant learning of new techniques.

4

Data Munging: String Manipulation, Regular Expressions, and Data Cleaning

This chapter is about some of the pathologies that you will see in real-world data. It talks about some of the most common (and notorious!) ones, where they come from, and how they can be addressed.

Data pathologies come in roughly two types. The first are formatting issues. This includes inconsistent capitalization, extraneous whitespaces, and things of that nature. Often, these are straightforward to solve with appropriate pre-processing of the data. The second category involves the actual content of the data. Duplicate entries, major outliers, and NULL values are all examples. It often requires some detective work to figure out what these issues mean in a particular situation and hence how they should be addressed.

My goals in this chapter are twofold. Firstly, I want to give you an appreciation for the breadth of issues that can be present in real-world data and equip you to quickly identify and diagnose problems. Secondly, I want to teach you tools that can be used to solve the problems. Specifically, I will discuss various types of string manipulation.

Manipulating strings of text might seem boring at first glance, but it's one of the most powerful tools a data scientist can have. I would put it on par with machine learning itself. String manipulation can be used to address any data formatting problems, and in many cases, it is the only suitable solution. But it is also invaluable for creating scripts to pull information out of raw data. Sometimes, when you encounter a new dataset, there is a "right" way to process it, which requires learning a new organizational paradigm and complicated tools that implement it. Alternatively, the quick-and-dirty way is to spend an hour hacking together a script that pulls out the specific data you need. You can guess which of these approaches is often more expedient if you need preliminary results by tomorrow.

The first part of this chapter will discuss a number of usual suspects when it comes to data issues. I will start with problems involving the data content, including some of the reasons they often arise. I will then move on to formatting issues and discuss how they can be addressed using strings. Finally, I will discuss the "big guns" in string manipulation: pattern matching via regular expressions.

4.1 The Worst Dataset in the World

The worst industrial dataset that I ever worked with was the first one. It was a collection of server logs, describing queries that had been received by a large collection of servers that my client owned. A given server could be referred to by a number of different names. Most lines of the logs were gobbledygook that were useless to me. Some of the key fields were encoded in weird hexadecimal. There were no rows or columns; instead, each line had its own structure. It was awful.

Then I worked with my second industrial dataset and discovered that they're all like that. Your worst dataset will probably be your first one too. Whenever there is a large organization, a complicated data collection process, or several datasets that have been merged, issues tend to pile up. They are rarely documented and often only come to light when some poor data scientist is tasked with analyzing them. You have been warned.

4.2 How to Identify Pathologies

One of the most embarrassing things that can happen in data science is to have to retract results that you've presented because you realize that you processed the data incorrectly. Given how convoluted datasets often are, you should have a healthy degree of paranoia about this happening.

To identify these issues early, I have four pieces of advice:

- If the data is text, look directly at the raw file rather than just reading it into your script.
- Read supporting documentation, if it's available. Often, the data is hard to understand because it uses strange codes or conventions, whose meaning is documented in some accompanying PDF files or something. In other cases, the data seems pretty self-explanatory, but there are nonobvious problems that only show up when you read the details.
- Have a battery of standard diagnostic questions you ask about the data. Does this column contain NULLs? Are all the identifiers in table A present in table B, and vice versa? Things like that.
- Do sanity checks, where you use the data to derive things that you already know. If you count the customers in the dataset and it isn't equal to the number of customers you know the company has, then chances are you weren't identifying unique customers correctly.

4.3 Problems with Data Content

4.3.1 Duplicate Entries

You should always check for duplicate entries in a dataset. Sometimes, they are important in some real-world way. In those cases, you usually want to condense them into one entry, adding an additional column that indicates how many unique entries there were.

In other cases, the duplication is purely a result of how the data was generated. For example, it might be derived by selecting several columns from a larger dataset, and there are no duplicates if you count the other columns.

4.3.2 Multiple Entries for a Single Entity

This case is a little more interesting than duplicate entries. Often, each real-world entity logically corresponds to one row in the dataset, but some entities are repeated multiple times with different data. The most common cause of this is that some of the entries are out of date, and only one row is currently correct.

In other cases, there actually should be duplicate entries. For example, each "entity" might be a power generator with several identical motors in it. Each motor could give its own status report, and all of them will be present in the data with the same serial number. Another field in the data might tell you which motor is actually which. In the cases where the motor isn't specified in a data field, the different rows will often come in a fixed order.

Another case where there can be multiple entries is if, for some reason, the same entity is occasionally processed twice by whatever gathered the data. This happens in many manufacturing settings, because they will retool broken components and send them through the assembly line multiple times rather than scrapping them outright.

4.3.3 Missing Entries

Most of the time when some entities are not described in a dataset, they have some common characteristics that kept them out. For example, let's say that there is a log of all transactions from the past year. We group the transactions by customer and add up the size of the transactions for each customer. This dataset will have only one row per customer, but any customer who had no transactions in the past year will be left out entirely. In a case such as this, you can join the derived data up against some known set of all customers and fill in the appropriate values for the ones who were missing.

In other cases, missing data arises because data was never gathered in the first place for some entities. For example, maybe two factories produce a particular product, but only one of them gathers this particular data about them.

4.3.4 NULLs

NULL entries typically mean that we don't know a particular piece of information about some entity. The question is: why?

Most simply, NULLs can arise because the data collection process was botched in some way. What this means depends on the context.

When it comes time to do analytics, NULLs cannot be processed by many algorithms. In these cases, it is often necessary to replace the missing values with some reasonable proxy. What you will see most often is that it is guessed from other data fields, or you simply plug in the mean of all the non-null values.

In other cases, the NULL values arise because that data was never collected. For example, some measurements might be taken at one factory that produces widgets but not at another. The table of all collected data for all widgets will then contain NULLs for whenever the widget's factory didn't collect that data. For this reason, whether a variable is NULL can sometimes be a very powerful feature. The factory that produced the widget is, after all, potentially a very important determinant for whatever it is you want to predict, independent of whatever other data you gathered.

4.3.5 Huge Outliers

Sometimes, a massive outlier in the data is there because there was truly an aberrant event. How to deal with that depends on the context.

Sometimes, the outliers should be filtered out of the dataset. In web traffic, for example, you are usually interested in predicting page views by humans. A huge spike in recorded traffic is likely to come from a bot attack, rather than any activities of humans.

In other cases, outliers just mean missing data. Some storage systems don't allow the explicit concept of a NULL value, so there is some predetermined value that signifies missing data. If many entries have identical, seemingly arbitrary values, then this might be what's happening.

4.3.6 Out-of-Date Data

In many databases, every row has a timestamp for when it was entered. When an entry is updated, it is not replaced in the dataset; instead, a new row is put in that has an up-to-date timestamp. For this reason, many datasets include entries that are no longer accurate and only useful if you are trying to reconstruct the history of the database.

4.3.7 Artificial Entries

Many industrial datasets have artificial entries that have been deliberately inserted into the real data. This is usually done for purposes of testing the software systems that process the data.

4.3.8 Irregular Spacings

Many datasets include measurements taken at regular spacings. For example, you could have the traffic to a website every hour or the temperature of a physical object measured at every inch. Most of the algorithms that process data such as this assume that the data points are equally spaced, which presents a major problem when they are irregular.

If the data is from sensors measuring something such as temperature, then typically you have to use interpolation techniques (which I discuss in a later chapter) to generate new values at a set of equally spaced points.

A special case of irregular spacings happens when two entries have identical timestamps but different numbers. This usually happens because the timestamps are only recorded to finite precision. If two measurements happen within the same minute, and time is only recorded up to the minute, then their timestamps will be identical.

4.4 Formatting Issues

4.4.1 Formatting Is Irregular between Different Tables/Columns

This happens a lot, typically because of how the data was stored in the first place. It is an especially big issue when joinable/groupable keys are irregularly formatted between different datasets.

4.4.2 Extra Whitespace

For such a small issue, it is almost comical how often random whitespace confounds analyses when people try to, say, join the identifier "ABC" against "ABC " for two different datasets. Whitespace is especially insidious because when you print the data to the screen to examine it, the whitespace might be impossible to discern.

In Python, every string object has a strip() method that removes whitespace from the front and end of a string. The methods lstrip() and rstrip() will remove whitespace only from the front and end, respectively. If you pass a character as an argument into the strip functions, only that character will be stripped. For example,

```
>>> "ABC\t".strip()
'ABC'
>>> "  ABC\t".lstrip()
'ABC\t'
>>> "  ABC\t".rstrip()
'  ABC'
>>> "ABC".strip("C")
'AB'
```

4.4.3 Irregular Capitalization

Python strings have lower() and upper() methods, which will return a copy of the original string with all letters set to uppercase or lowercase.

4.4.4 Inconsistent Delimiters

Usually, a dataset will have a single delimiter, but sometimes, different tables will use different ones. The most common delimiters you will see are as follows:

- Commas
- Tabs
- Pipes (the vertical line "|").

4.4.5 Irregular NULL Format

There are a number of different ways that missing entries are encoded into CSV files, and they should all be interpreted as NULLs when the data is read in. Some popular examples are the empty string "", "NA," and "NULL." Occasionally, you will see others such as "unavailable" or "unknown" as well.

4.4.6 Invalid Characters

Some data files will randomly have invalid bytes in the middle of them. Some programs will throw an error if you try to open up anything that isn't valid text. In these cases, you may have to filter out the invalid bytes.

The following Python code will create a string called s, which is not validly formatted text. The decode() method takes in two arguments. The first is the text format that the string should be coerced into (there are several, which I will discuss later in the chapter on file formats). The second is what should be done when such coercion isn't possible; saying "ignore" means that invalid characters simply get dropped.

```
>>> s = "abc\xFF"
>>> print s # Note how last character isn't a letter
abc□
>>> s.decode("ascii", "ignore")
u'abc'
```

4.4.7 Weird or Incompatible Datetimes

Datetimes are one of the most frequently mangled types of data field. Some of the date formats you will see are as follows:

- August 1, 2013
- AUG 1, '13
- 2013-08-13

There is an important way that dates and times are different from other formatting issues. Most of the time you have two different ways of expressing the same information, and a perfect translation is possible from the one to the other. But with dates and times, the information content itself can be different. For example, you might have just the date, or there could also be a time associated with it. If there *is* a time, does it go out to the minute, hour, second, or something else? What about time zones?

Most scripting languages include some kind of built-in datetime data structure, which lets you specify any of these different parameters (and uses reasonable defaults if you don't specify). Generally speaking, the best way to approach datetime data is to get it into the built-in data types as quickly as possible, so that you can stop worrying about string formatting.

The easiest way to parse dates in Python is with a package called dateutil, which works as follows:

```
>>> import dateutil.parser as p
>>> p.parse("August 13, 1985")
datetime.datetime(1985, 8, 13, 0, 0)
>>> p.parse("2013-8-13")
datetime.datetime(2013, 8, 13, 0, 0)
>>> p.parse("2013-8-13 4:15am")
datetime.datetime(2013, 8, 13, 4, 15)
```

It takes in a string, uses some reasonable rules to determine how that string is encoding dates and times, and coerces it into the datetime data type. Note that it rounds down – August 13th becomes 12:00 AM on August 13th, and so on.

4.4.8 Operating System Incompatibilities

Different operating systems have different file conventions, and sometimes, that is a problem when opening a file that was generated on one OS on a computer that runs a different one.

Probably, the most notable place where this occurs is newlines in text files. In Mac and Linux, a newline is conventionally denoted by the single character "\n." On Windows, it is often two characters "\r\n." Many data processing tools check what operating system they are being run on so that they know which convention to use.

4.4.9 Wrong Software Versions

Sometimes, you will have a file of a format that is designed to be handled by a specific software package. However, when you try to open it, a very mystifying error is thrown. This happens, for example, with data compression formats.

Oftentimes the culprit ends up being that the file was originally generated with one version of the software. However, the software has changed in the meantime, and you are now trying to open the file with a different version.

4.5 Example Formatting Script

The following script illustrates how you can use hacked-together string formatting to clean up disgusting data and load it into a Pandas DataFrame. Let's say we have the following data in a file:

```
Name|Age|Birthdate
Ms. Janice Joplin|65|January 19, 1943
  Bob Dylan |74 Years| may 24 1941
Billy Ray Joel|66yo|Feb. 9, 1941
```

It's clear to a human looking at the data what it's supposed to mean, but it's the kind of thing that might be terrible if you opened it with a CSV file reader. The following code will take care of the pathologies and make things more explicit. It's not exactly pretty or efficient, but it gets the job done, it's easy to understand, and it would be easy to modify if it needed changing:

```python
def get_first_last_name(s):
    INVALID_NAME_PARTS = ["mr", "ms", "mrs",
        "dr", "jr", "sir"]
    parts = s.lower().replace(".","").strip().split()
    parts = [p for p in parts
        if p not in INVALID_NAME_PARTS]
    if len(parts)==0:
        raise ValueError(
            "Name %s is formatted wrong" % s)
    first, last = parts[0], parts[-1]
    first = first[0].upper() + first[1:]
    last = last[0].upper() + last[1:]
    return first, last

def format_age(s):
    chars = list(s) # list of characters
    digit_chars = [c for c in chars if c.isdigit()]
    return int("".join(digit_chars))

def format_date(s):
    MONTH_MAP = {
```

```
        "jan": "01", "feb": "02", "may": "03"}
    s = s.strip().lower().replace(",", "")
    m, d, y = s.split()
    if len(y) == 2: y = "19" + y
    if len(d) == 1: d = "0" + d
    return y + "-" + MONTH_MAP[m[:3]] + "-" + d
import pandas as pd
df = pd.read_csv("file.tsv", sep="|")
df["First Name"] = df["Name"].apply(
    lambda s: get_first_last_name(s)[0])
df["Last Name"] = df["Name"].apply(
    lambda s: get_first_last_name(s)[1])
df["Age"] = df["Age"].apply(format_age)
df["Birthdate"] = df["Birthdate"].apply(
    format_date).astype(pd.datetime)
print df
```

4.6 Regular Expressions

Regular expressions are one of the "big guns" standard tools in data processing. They take many of the operations we just discussed (split, index, etc.), which take in specific strings as arguments, and generalize them to apply to a pattern. For example, say we want to pull all phone numbers that match the (XXX) XXX-XXXX pattern out of a document. That would be very onerous with normal strings but a synch with regular expressions. The "regular expression" is a string that encodes the pattern you're looking for.

Before we go much farther, I should let you know that regular expressions are a bit notorious for being finicky to use and debug. This is because while it is possible to express a stupendous array of patterns within regular expressions, the expressions themselves quickly become complicated enough that it's hard for humans to wrap their heads around. This is a fairly fundamental problem – "understanding" a pattern you want is much, much easier than specifying every jot and tittle of what constitutes that pattern. I'm reminded of the phrase "damn it computer: do what I *want*, not what I say."

The way around this is to avoid regular expressions that are overly complex. So as long as you keep them short enough that they are easy to understand, they are extremely powerful.

The other caveat about regular expression is that they are computationally expensive. There are many different ways that a piece of text can potentially match a complicated pattern, and it takes a while to check them all. In fact, even the process of compiling the regular expression itself into a computation-ready data structure takes a while.

4.6.1 Regular Expression Syntax

Let's start with a very simple regular expression: "ab*". This means that the pattern is exactly one occurrence of the letter "a," followed by some number (possibly 0) of "b"s – the "*" means arbitrary repetition of whatever letter came right before it. In the string "abcd abb," the pattern occurs twice: the initial "ab" and the "abb" at the end.

Right there though we run into the first subtle point of regular expressions: how do we pick which matches to find? We said that the "ab" at the start was a match, because it was an "a" followed by one "b". But technically the "a" would have been a match on its own – it's just a match with zero "b"s rather than one. In situations such as this, do we want to find all possible matches even if they overlap? If we only want nonoverlapping matches, how do we pick which ones? The general answer is to start at the left-most part of the text and find the largest possible match. Then you chop this match out of the text and find the largest possible match from the remaining text, and so on. This is called the "greedy" approach – there are times when you want to override it, but they're typically not common.

Greedy parsing is fantastic from a performance perspective, because it requires a single pass through a piece of text and generally only requires a small portion of it to be kept in memory at any one time. Regular expression is still inefficient, but this does much to ease the pain. In many applications, the matches will be returned as they are found, with the whole parser acting as one giant iterator. Especially for large pieces of text where many matches are expected, this is the most efficient way to do it. The Match objects returned by the iterator will include several pieces of information such as the matching string itself and its start/end indices. However, in my own work, I usually end up using simpler approaches that just return all matches as a list of strings, rather than an iterator of Match objects.

Now let's see some of the more complicated types of pattern that can be specified:

Type of pattern	Regular expression	Example matches	Notes
A fixed string	abc123	abc123	"abc123" contains no special characters, so it's just a string to be matched.
Arbitrary repetition	a*b	b ab aaab	"*" means that you can have an arbitrary number (possibly 0) of the previous character.
Repeat character at least once	a+b	ab aaaab	

Type of pattern	Regular expression	Example matches	Notes
Repeat character at most once	a?b	b ab	
Repeat a character a fixed number of times	a{5}	aaaaa	
Repeat a pattern a fixed number of times	(a*b){3}	baabab ababaaaab	
Repeat a character or pattern a variable number of times	a{2,4}	aa aaa aaaa	Note that the range is inclusive.
Choice of several characters	[ab]c	ac bc	The brackets means that you can have any single character from within the brackets.
Arbitrary mixture of several characters	[ab]*c	c aac abbac	In this case, the * is applied to the whole [ab] expression.
Ranges of characters	[A-H][a-z]*	Aasdfalsd Hb G	[A-H] is shorthand for the characters from A to H. You can do the same thing with digits.
Characters OTHER than a particular one	[^AB]	C D	The ^ as the first argument in [] means to match any character NOT in that group. If ^ is not the first character, then it has no special meaning.
Choice of several expressions	Dr|Mr|Ms|Mrs	Dr Mr Ms Mrs	Here you can select.
Nesting expressions	([A-Z]|[a-z]|[0-9])*	A AzSDFcvfg	This matches any alphanumeric string. In Python, \w is shorthand for this.
Start of a line	^ab		Matches any "ab" that occurs at the start of your text or just after a newline.
End of a line	ab$		Matches an "ab" that is at the end of the document or just after a newline.

(Continued)

Type of pattern	Regular expression	Example matches	Notes
Special characters	\[	[	If you want to include one of the special characters in your pattern, you can escape it with a \.
Any character except newline	.	a * —	
Nongreedy evaluation	<.*>?	\<h1> </h2 name="foo">	This causes it to find the shortest possible path, rather than the longest.
Whitespace	\s		This matches any whitespace character, such as spaces, tabs, or a newline

There are others, and regular expression syntax can vary a little bit between languages and libraries. However, these should be enough to get you started.

The nongreedy character deserves some special explanation. Let's say that we have the following XML data:

`<name>Jane</name><name>Bob</name>`

and we want to pull out the name fields. You might try to use the regular expression

`<name>.*</name>`

but that will end up matching the entire string. This is because "</ name><name" in the middle matches the ".*", and regular expression try to match as much text as possible. If instead you say

`<name>.*?</name>`

then you will get the two matches, because it tries to match ".*" to as little text as possible.

Python's implementation is a relatively lightweight library called re. The following Python code shows how to read in a file and use regular expressions to look for street addresses. It's not perfect, but it will work pretty well.

```
import re
# This matches "1600 Pennsylvania Ave."
# It does NOT match "5 Stony Brook St"
# cuz there is a space in "Stony Brook"
street_pattern = r"^[0-9]\s[A-Z][a-z]*" + \
    r"(Street|St|Rd|Road|Ave|Avenue|Blvd|Way|Wy)\.?$"
# Like the one above, this assumes
# there is no space in the town name
```

```
city_pattern = r"^[A-Z][a-z]*,\s[A-Z]{2},[0-9]{5}$"
address_pattern = street_pattern + r"\n" \
    + city_pattern
# Compile the string into a regular expression object
address_re = re.compile(address_pattern)
text = open("some_file.txt", "r").read()
matches = re.findall(address_re, text)
# list of all strings that match
open("addresses_w_space_between.txt",
    "w").write("\n\n".join(matches))
```

You should notice the following things about that code:

1) It's very powerful! This is only a few lines, but it is doing a very complicated task.

2) It's limited. There are many idiosyncrasies of addresses that the human eye can spot that will elude this regular expression. It won't handle apartment numbers, multiword street names, or even "32nd street." You can patch these problems up as you find them, but you risk the code becoming unwieldy.

3) We are declaring our strings as "raw strings," by putting an r in front of the opening quote.

The last thing is a practical measure when doing regular expressions in Python, because using the escape character \ can become a massive pain. The problem is that if we say

```
pattern = "\n"
my_re = re.compile(pattern) # trying to match a newline
```

we have not done what we intended to do. The string called pattern is a one-character string, consisting of the newline character. But re.compile would require a two-character string, with the first character being a slash and the second being an n. We could instead have said

```
# Escape the slash w another
slash pattern = "\\n"
# This matches a newline
newline_re = re.compile(pattern)
```

But this becomes extremely unwieldy if we want to, say, include the slash character in the pattern we are looking for. The pattern to match a single slash would be "\\\\".

Putting the r before the quotes in Python creates a "raw string," meaning that the exact contents of the quotes is the string. Life is just easier that way.

4.7 Life in the Trenches

This is a fairly short chapter, because there isn't a lot to say about data cleaning that generalizes well. In a lot of ways, it is the boring part of data science, a price we must pay to get things into a format where we can ask the real questions.

At the same time though, it is an intellectually challenging, problem-solving activity – often more so than the analysis itself. In many data science projects, there is a staggering amount of detective work and coding required to just get the data into clean tabular form, but after that all you do is fit a line or something equally trivial.

Data cleaning code is one of the areas where data science blurs into production coding. There is a lot of room for creativity and experimentation in how you extract features from data or what analyses you run. Generally though, there is only one "right" way to clean the data, and the code tends to get written once and then reused between different iterations of analysis and feature extraction.

Once you understand the data itself and have written the cleaning scripts, it is time to move on to understanding the world it is describing. This is the world of visualizations and exploratory analysis.

4.8 Glossary

Regular expression A way to specific a general pattern that strings can match. Regular expressions can be finicky to use, but they are extremely powerful.

String formatting A nifty way in Python and many other languages to insert content into template strings.

5

Visualizations and Simple Metrics

A rule of thumb for data science deliverables is this: if there isn't a picture, then you're doing it wrong. Typically, a good analytics project starts (after cleaning and understanding the data) with exploratory visualizations that help you develop hypotheses and get a feel for the data, and it ends with carefully manicured figures that make the final results visually obvious. The actual number crunching is hidden in the middle, sometimes almost as an aside. I've had a number of projects where there was never even any actual machine learning: people needed to know whether there was signal in the data and which directions were most promising for further work (which would potentially include machine learning), and graphics showed that more clearly than a number ever could.

This fact is very underappreciated outside of the data analysis community. Many people think of data scientists as numerical badasses, working black magic from a command line. But that's just not the way the human brain processes data, generates hypotheses, or develops familiarity with an area. Pictures are plans A–C for everything except the last stages of statistically validating results. I've often joked that if humans were able to visualize things in a thousand dimensions, then my job as a data scientist would consist entirely of generating and looking at scatterplots.

This chapter will take you through several of the most important visualizations. You've probably seen most of this before, but it's always good to revisit the basics. We will also cover some exploratory metrics (such as correlations), which capture, in crude numerical form, some of the patterns that are clear from a good visual. There are many techniques not covered in this chapter, and you would do well to learn them. However, my experience is that these core ones will cover most of your needs. I strongly recommend memorizing the syntax for basic visualizations in your programming language of choice. In exploratory analysis especially, it's useful to be able to chug through various ways of visualizing your data without needing to consult a reference on the syntax.

The Data Science Handbook, First Edition. Field Cady.
© 2017 John Wiley & Sons, Inc. Published 2017 by John Wiley & Sons, Inc.

There are, however, still times when we need a number. There are two reasons for this:

- Our eyes can trick us, so it's important to have a cold hard statistic too.
- Often, you don't have time to sift through every possible picture, and you need some way to put a number on it so that the computer can make decisions of some sort automatically (even if the decision is only which pictures are worth your time to look at).

Besides visualization techniques, this chapter will cover some standard statistical metrics that strive to capture, in numerical form, some of the meaning that you can get out of a picture.

5.1 A Note on Python's Visualization Tools

The main visualization tool for Python is a library called matplotlib. While matplotlib is powerful and flexible, it is probably the weakest link in Python's technical stack. The graphs can be a bit cartoonish, in some ways the syntax is nonintuitive, and the interactivity (zooming in, etc.) leaves something to be desired. Most of the appearance issues can be fixed by tweaking a graphic's configuration, but the default settings are not great.

I'm sticking with matplotlib for this book because it is by far the most standard tool, it is sufficient for most data science (especially if you learn some of the ways you can make the plots look prettier), and it integrates well with the other libraries. But there are other libraries out there that are gaining ground, especially browser-based ones such as Bokeh and Plot.ly.

Example code in this chapter will use Pandas whenever possible. However, Pandas' visualizations are a wrapper-around matplotlib, and sometimes, we have to use matplotlib directly. Typically, you make an image by calling the plot() method on a Pandas object, and Pandas does all the image formatting under the hood. Then you use matplotlib's pyplot module for things such as setting the title and the final act of either displaying the image or saving it to a file.

5.2 Example Code

To illustrate the visualization techniques we discuss in this chapter, we will apply them to the famous Iris dataset, which you may have seen in a statistics textbook. It describes physical measurements taken of flower specimens, drawn from three different species of iris. There are 150 data points, 50 from each species, and each data point gives the length and width of the pedals and sepals.

The following code sets the stage for all of the example code in this chapter. It imports the relevant libraries and creates a DataFrame containing the sample dataset (which comes built-in to scikit-learn):

```python
import pandas as pd
from matplotlib import pyplot as plt
import sklearn.datasets
def get_iris_df():
  ds = sklearn.datasets.load_iris()
  df = pd.DataFrame(ds['data'],
    columns = ds['feature_names'])
  code_species_map = dict(zip(
    range(3), ds['target_names']))
  df['species'] = [code_species_map[c]
    for c in ds['target']]
  return df
df = get_iris_df()
```

5.3 Pie Charts

Pity the poor pie chart. I feel like I never see it used in "serious" applications, almost as if it's looked down on as being too simple. But pie charts are really one of the clearest ways to present data, and I recommend using them whenever they're applicable. Technically, everything you get from a pie chart you could get equally well from looking at a list of numbers, but making sense of the numbers requires cognitive effort and attention. On the other hand, lower-level neural circuits make immediate sense of pie charts. This is perhaps the clearest illustration of the guiding principle behind all visualizations: it's not about conveying information, but about conveying it in a way that the human brain will understand and care about.

In my own work, the most mileage I've gotten out of pie charts is when I'm either doing exploratory analysis of a dataset (How many of our customers are senior citizens? How many of the page views came from the United States?) or communicating the results of a binary classifier.

The code to generate a basic pie chart using Pandas is very simple:

```python
sums_by_species = df.groupby('species').sum()
var = 'sepal width (cm)'
sums_by_species[var].plot(kind='pie', fontsize=20)
plt.ylabel(var, horizontalalignment='left')
plt.title('Breakdown for ' + var, fontsize=25)
plt.savefig('iris_pie_for_one_variable.jpg')
plt.close()
```

It will produce this figure:

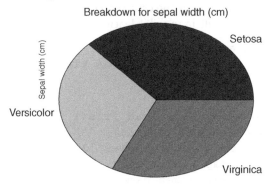

Note that some of the text overlaps. Little things such as this can happen in matplotlib if you use the default settings, and, in general, you will need to tweak the graph's configurations if you want things to look polished.

The previous figure was made by calling the plot() method on a Pandas Series object, whose index gave the flower species. If we instead call it on a DataFrame with multiple columns, we can generate a different chart for each column all in the same figure:

```
sums_by_species = df.groupby('species').sum()
sums_by_species.plot(kind='pie', subplots=True,
layout=(2,2), legend=False)
plt.title('Total Measurements, by Species')
plt.savefig('iris_pie_for_each_variable.jpg')
plt.close()
```

That code will give us the following:

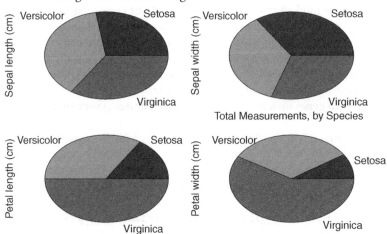

5.4   Bar Charts

The same information that is in a pie chart could equally well be conveyed in a bar chart. In this particular case, it's actually a much more sensible visualization, since we're interested in the relative sizes of the different flowers rather than how big a slice of the "flower pie" they each take up. The following code

```
sums_by_species = df.groupby('species').sum()
var = 'sepal width (cm)'
sums_by_species[var].plot(kind='bar', fontsize=15,
rot=30)
plt.title('Breakdown for ' + var, fontsize=20)
plt.savefig('iris_bar_for_one_variable.jpg')
plt.close()
sums_by_species = df.groupby('species').sum()
sums_by_species.plot(
    kind='bar', subplots=True, fontsize=12)
plt.suptitle('Total Measurements, by Species')
plt.savefig('iris_bar_for_each_variable.jpg')
plt.close()
```

will produce the following visualizations:

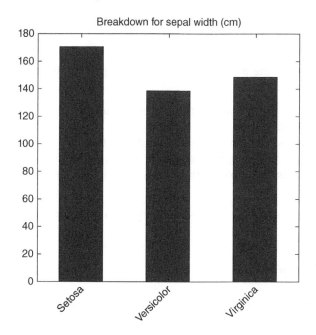

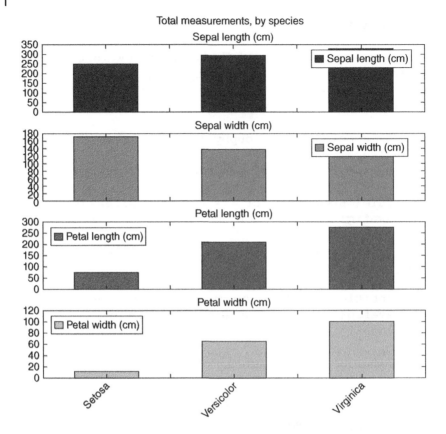

Note the following pieces of Python's plotting syntax that let us tweak the appearance of the figure:

- The "font size" optional argument controls how big a piece of font is. It's usually too small by default.
- The "rot" optional argument lets us rotate the text.
- We used suptitle() to give the title for the overall figure – Pandas will by default label each subplot with its corresponding column in the DataFrame being plotted.

There are others available if you are trying to make the figures look really polished.

5.5 Histograms

Histograms are probably my personal favorite visualization tool, partly because it seems like they usually contain something interesting. There are often distinct bumps in the histogram, which might correspond to several distinct classes of real-world entities. You can get a sense of whether there are a few distinct outliers, how much variation is in the population, and so on. A histogram is almost always a meaningful thing to make; it works for floating values, or integers, and unlike scatterplots, you only need one numerical field.

The following code will produce histograms for all the columns and put them together in one figure:

```
df.plot(kind='hist', subplots=True, layout=(2,2))
plt.suptitle('Iris Histograms', fontsize=20)
plt.show()
```

The final image appears as follows:

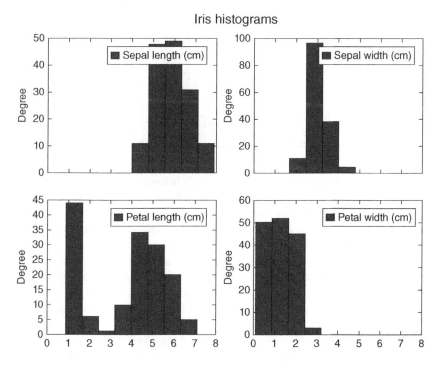

What jumps out at you is that the petal length has a clearly bimodal distribution, suggesting that this one species has almost categorically longer petals. We can confirm this by plotting each species separately, but on the same axes and in different colors:

```
for spec in df['species'].unique():
  forspec = df[df['species']==spec]
  forspec['petal length (cm)'].plot(
    kind='hist', alpha=0.4, label=spec)
```

```
plt.legend(loc='upper right')
plt.suptitle('Petal Length by Species')
plt.savefig('iris_hist_by_spec.jpg')
```

It yields the following graph. I know you can't see the color in the book you're holding, but sure enough the peak on the left is only the iris setosa variety.

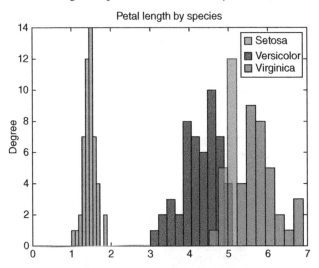

There are two big problems that occur with histograms. The first one is the number and size of the bins you use. If the bins are too large, then you can obscure fascinating patterns that occur within a single bucket. If they are too small, then many of your buckets will contain no points, and your bell-shaped curve will turn into a bunch of one-unit high bars.

The second problem is that sometimes your data can mar the picture. There might be one bucket that contains so many points, for example, that every other bucket is squashed down to what looks like noise. There might, for example, be a massive spike at 0.0, and you have to filter out those points before you draw the histogram.

The other visual problem is outliers, which can smash the overwhelming majority of the points to the far left of the graph. In some cases, this is pretty simple to deal with – you have a handful of points that are massive outliers, and all you need to do is filter out those points. Those points are aberrations, and it makes sense to remove them before drawing analytical conclusions.

But in my experience, it's usually not that simple. Rather than a handful of massive outliers, there is often a fat tail, representing a very real phenomenon in your data. You can cut part of the tail out of your dataset for purposes of making the visualization clearer, and you will probably have to, but in doing so, you will be cutting out very real, meaningful signal. You are not throwing out a few points that are clearly aberrations; you're picking a more-or-less arbitrary threshold and looking only at the part of your dataset that falls below it.

5.6 Means, Standard Deviations, Medians, and Quantiles

Sometimes, of course, you must summarize a distribution down to just a few numbers. Usually, these summaries are based on the assumption that your data's distribution is bell-shaped, and your goal is to give some idea of where the peak of the bell is and how widely it spreads. Within this vein, there are two main options:

1) Give the mean and standard deviation. These are the more historically popular metrics, and they are much easier to compute.
2) Give the median, 25th percentile, and 75th percentile. These metrics are more robust to pathologies in the data, but they are computationally more expensive (since you must sort a list).

They can be calculated as follows:

```
col = df['petal length (cm)']
Average = col.mean()
Std = col.std()
Median = col.quantile(0.5)
Percentile25 = col.quantile(0.25)
Percentile75 = col.quantile(0.75)
```

These numbers all still exist even if your data has multiple peaks in the distribution, but their usual intuitive interpretation breaks down.

The other pathology that deserves some discussion is outliers in the data. This isn't a huge problem for medians and quantiles, but it can be game-changing with mean and standard deviation. This is shown in the following figure, where I have simulated data from a log-normal distribution (log-normals are prone to outliers), made a histogram, and plotted the mean as a dashed vertical line. The handful of very large outliers have pulled the mean well to the right of the actual hump in the distribution.

Outliers make it very hard to give an intuitive interpretation of the mean, but in fact, the situation is even worse than that. For a real-world distribution, there always *is* a mean (strictly speaking, you can define distributions with no mean, but they're not realistic), and when we take the average of our data points, we are trying to estimate that mean. But when there are massive outliers, just a single data point is likely to dominate the value of the mean and standard deviation, so much more data is required to even *estimate* the mean, let alone make sense of it.

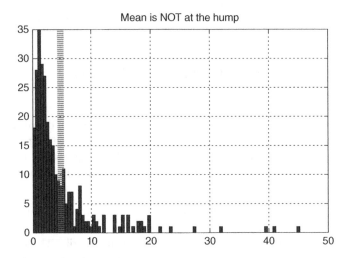

A common solution is to, before calculating the mean, throw out all data points that are deemed to be outliers: a common criterion is anything below the 25th percentile or above the 75th percentile. You can do that as follows:

```
col = df['petal length (cm)']
Perc25 = col.quantile(0.25)
Perc75 = col.quantile(0.75)
Clean_Avg = col[(col>Perc25)&(col<Perc75)].mean()
```

This workaround corresponds to the idea that these outliers are pathological data points, which really *should* be discarded if we're trying to understand the underlying phenomena. If we're dealing with measurements from a physical sensor, for example, they might have been caused by a malfunction of our hardware. In other situations though, such as the amount of money in a transaction, the outliers are extremely important data points that can't be discarded.

The median is not perfect either. In the case of outlier data, or even just a lopsided bell curve, it moves away from the hump in the bell curve. The median also does not change *at all* if you perturb the outlier values. However, it still keeps its user-friendly meaning: half the values are greater and half are less.

Personally, I generally use median if I want to know what's "typical," but I use mean if the average behavior is really what I care about from a business perspective.

5.7 Boxplots

Boxplots are a convenient way to summarize a dataset by showing the median, quantiles, and min/max values for each of the variables. The following code snippet makes a boxplot of the sepal length for each of the species in the iris dataset. The boxplot makes it glaringly obvious that the three species are different from one another.

```
col = 'sepal length (cm)'
df['ind'] = pd.Series(df.index).apply(lambda i: i% 50)
df.pivot('ind','species')[col].plot(kind='box')
plt.show()
```

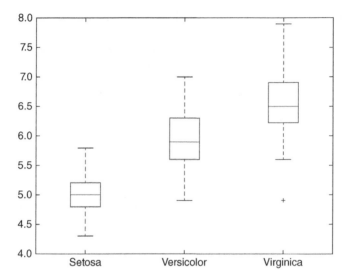

An advantage of boxplots is that major outliers are very visually obvious. Here is a boxplot of the data we put in a histogram in the previous section:

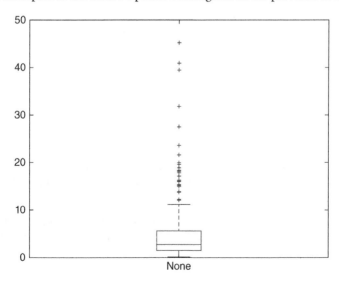

Note that the upper quantile is much farther from the median compared to the lower quantile, and the effect is even more pronounced for the min and max values. If you just use the histogram, outliers can show up as a deceptively slight increase in the thickness of the tails.

5.8 Scatterplots

I've often joked that if humans could see things in an arbitrary number of dimensions, then all of my data science work would consist of making and interpreting scatterplots. In my experience, they are one of the simplest but most powerful ways to visualize relationships within a dataset, so they're a great first step when you're finding your feet with a new project.

A simple scatterplot is very easy to generate in Python:

```
df.plot(kind="scatter",
    x="sepal length (cm)", y="sepal width (cm)")
plt.title("Length vs Width")
plt.show()
```

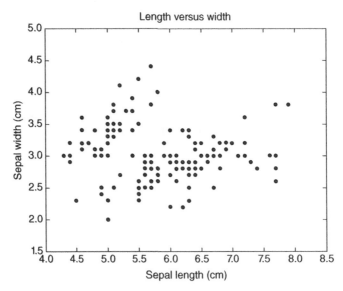

Besides the basic plotting, scatterplots can have several other bells and whistles, which allow more than just the two dimensions to be packed in. They include the following:

- Color coding. Often data points that fall into different categories are given different colors. You can also use color in a continuous way, such as different mixtures of red and blue.

- Size. Changing the size of data points communicates another dimension of information, similarly to color coding. It also has the often-desirable ability to draw attention disproportionately to some points instead of others.
- Opacity. In scatterplots and other visualizations, it is often useful to make things partially transparent in case they overlap with other parts of the visualization.

These parameters are often useful when doing exploratory analysis of a dataset, but they can be especially compelling when you're putting together final visualizations for use in final reports and presentations.

If you want to control the formatting of a plot in Python, you can do it by passing optional arguments into the scatter() function. The ones of most interest are as follows:

- c: A string indicating the color to make the dots. You can also pass in a sequence of such strings if the dots are to be of different colors.
- s: The size that each point should be, in pixels. Alternatively, you can pass in a sequence of sizes.
- marker: A string indicating what marker should be used in the plot.
- alpha: The transparency.

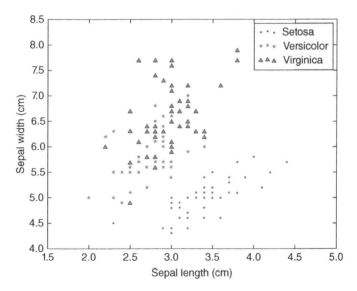

For example, the following script will produce the figure indicated:

```
plt.close()
colors = ["r", "g", "b"]
markers= [".", "*", "^"]
```

```
fig, ax = plt.subplots(1, 1)
for i, spec in enumerate(df['species'].unique() ):
  ddf = df[df['species']==spec]
  ddf.plot(kind="scatter",
    x="sepal width (cm)", y="sepal length (cm)",
    alpha=0.5, s=10*(i+1), ax=ax,
    color=colors[i], marker=markers[i], label=spec)
plt.legend()
plt.show()
```

It is immediately clear from the picture that the iris setosa flowers stand apart as having sepals that are markedly longer and narrower.

5.9 Scatterplots with Logarithmic Axes

A key variation on scatterplots is using logarithmic axes. In many applications, the numbers being plotted are all positive (or at least nonnegative), but they can vary by orders of magnitude. This might happen if you are looking at traffic to a collection of websites, where some sites receive vastly more views than others or personal income. In a scatterplot of data such as this, all but the largest data points will be squashed to one side, making the plot essentially unreadable.

Here is a good example using a dataset from scikit-learn. I am making a scatterplot of the crime rate in a neighborhood versus the median home value:

```
import pandas as pd
import sklearn.datasets as ds
import matplotlib.pyplot as plt
# Make Pandas dataframe
bs = ds.load_boston()
df = pd.DataFrame(bs.data, columns=bs.feature_names)
df['MEDV'] = bs.target
# Normal Scatterplot
df.plot(x='CRIM',y='MEDV',kind='scatter')
plt.title('Crime rate on normal axis')
plt.show()
```

Note how almost all of the data points are squashed to the left, making the graph hard to read for all but the most high-crime neighborhoods.

Instead, we could have made the *x*-axis logarithmic, as follows:

```
df.plot(x='CRIM',y='MEDV',kind='scatter',logx=True)
plt.title('Crime rate on logarithmic axis')
plt.show()
```

The two code snippets will create the following scatterplots

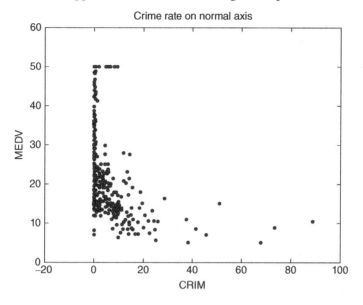

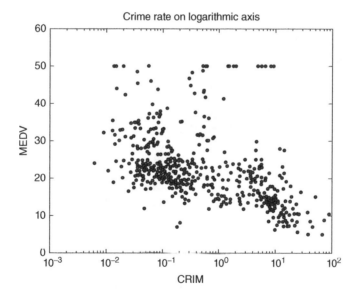

In the second plot the tick marks on the *x*-axis are irregular. They are all equally spaced in the figure. However, they correspond to small changes in numbers on the left but large numbers on the right. This has the effect of taking our original graph and widening out the left-hand part while we shorten out the right-hand part.

With this rescaling, we can see that there is a clear inverse relationship between crime rate and median home value that exists across all levels of crime.

Mathematically, we make a logarithmic plot by taking the log of the raw data and using that to tell where to place the points. The caveat to this is that logarithmic plots only work when all values are greater than 0. In many situations, for example, if you are counting events, the data can be 0 but is guaranteed to never be negative. In this case, it's common to just add 1 to the data and plot that instead. But in situations where the data can be arbitrarily negative, logarithmic plots are not appropriate.

5.10   Scatter Matrices

The biggest problem with scatterplots is that we often have many different variables to compare, and human visualization abilities top out at three dimensions. A partial solution to this is to do a scatterplot comparing every pair of features, arranging them in what's sometimes called a "scatter matrix" as shown here:

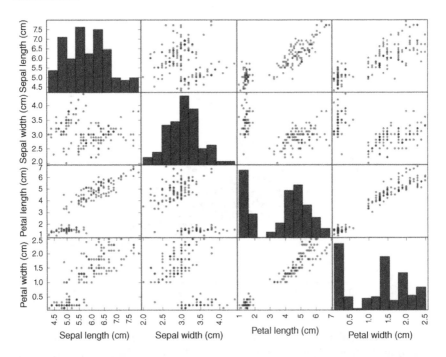

Note that along the diagonal, we have a histogram of each feature, rather than a scatterplot of the feature versus itself (which would just be a straight line).

The code that generated that visual is

```
plt.close()
from pandas.tools.plotting import scatter_matrix
scatter_matrix(df)
plt.show()
```

5.11   Heatmaps

Another problem with scatterplots is that they can become visually cluttered if you have a lot of data points. You can ameliorate the problem by reducing the size of the data points (maybe a good idea anyway – default point sizes are often annoyingly big), but that only goes so far. It's an especially useless workaround if many of your points are exactly on top of each other, which can easily happen if your data is integers rather than floats. Eventually, your scatterplot becomes just a big mass of overlapping points with no background visible, and there is no way to tell which areas have more or fewer points. It is possible to use the alpha parameter to adjust the transparency of the points, so that they become darker when there is more overlap, but this becomes very clunky.

In that situation, we don't actually care about the actual points themselves. We care about the density of points in the different regions, and the correct

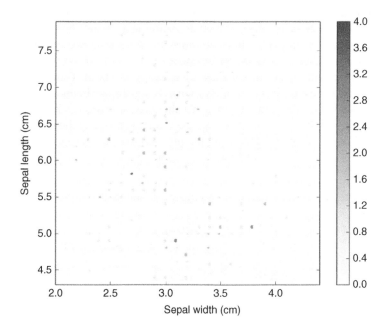

way to visualize that is with a heatmap, which color-codes different regions of the plane by their relative density of points. In some applications (including Pandas), those regions are small hexagons, and they are called "hexbin" heatmaps.

The code for generating a heatmap, along with the resulting figure, is

```
plt.close()
df.plot(kind="hexbin",
    x="sepal width (cm)", y="sepal length (cm)")
plt.show()
```

5.12 Correlations

If you hear the term "correlation" used casually, it is probably what's called "Pearson" correlation. More generally, a correlation is a metric that measures how closely tied two variables X and Y are, and there are two main types you'll see:

1) Pearson correlation. This is the normal one, and it measures how accurate it is to say that

$$Y = mx + b$$

   A correlation near 1 means that, for some b and some m>0, this equation is a good approximation. If the correlation is near −1, then it means the same thing, except m is negative. Note that assuming a linear relationship is very restrictive. If $Y = Sqrt[X]$, they still move up/down together exactly in sync, but they will have a correlation less than 1.

2) Ordinal correlations. This makes no assumption about X and Y having a linear relationship. It just models their relationship as being monotonic: if you sort your data points by their X value (say, somebody's height), is that more or less the same order you get from sorting them by their Y value (say, somebody's weight)? Do the taller people tend to be the heavier ones? There are two main types of ordinal correlation that you'll see: Spearman and Kendall.

The correlations in Pandas can be simply calculated as follows:

```
>>> df["sepal width (cm)"].corr(
    df["sepal length (cm)"])   # Pearson corr
-0.10936924995064937
>>> df["sepal width (cm)"].corr(
    df["sepal length (cm)"], method="pearson")
-0.10936924995064937
```

```
>>> df["sepal width (cm)"].corr(
    df["sepal length (cm)"], method="spearman")
-0.15945651848582867
>>> df["sepal width (cm)"].corr(
    df["sepal length (cm)"], method="spearman")
-0.072111919839430924
```

None of the correlations measure how related two variables are. If $Y = \sin(X)$, for example, and X covers a wide range, then in some cases they go up together and in other cases they go down, and will have a correlation near 0. The best way to try and correct for this is to plot their relationship.

The two ordinal correlations are similar to each other, and usually, either will work. In general though, Kendall correlation is more robust to aberrant data points, for example, if the tallest person in a room was also the *least* heavy. Conversely, Kendall is very sensitive to small changes in ordering that are often inconsequential for a real application: if the tallest person in the room is only the *second* heaviest, Kendall will punish that deviation much more severely compared to Spearman. If I have to choose, I usually use Spearman.

I periodically get asked about how strong a correlation needs to be to be "strong enough." The answer to this depends entirely on context, and I can't give you any absolute rules. Personally, I start to care when the absolute value of the correlation gets above 0.4. Around 0.7, we're starting to talk about using one variable to defensibly estimate the other – there is a rigorous sense in which "half of the variation" of the one variable can be explained by the other at this level of correlation. Anything over 0.95 and I figure that one variable is basically a synonym for the other, and there was probably some weirdness in the data that caused this.

Finally, you've probably heard a lot about how "correlation is not causation." The gold standard in science is controlled experiments, where we forcibly change one (and only one) experimental parameter and then see how other things change. Reliable controlled experiments are, strictly speaking, the only safe way to conclude that one thing causes another. But especially in areas such as sociology or economics, this is usually impossible, and if all we know is that two things are correlated in the real world, we cannot rigorously conclude anything about causality.

If thing A and thing B are highly correlated, then humans have an almost pathological need to say that A causes B (or vice versa – whichever one sounds more plausible). Usually, neither is really true, and I would love for my personal contribution to the lore of statistics to be the following:

> *Cady's Rule of Thumb*: If A and B are correlated, then neither one is causing the other. Instead, there is some factor C causing them both.

If thing A and thing B are correlated, then try to figure out what C might be. This is a great way to generate hypotheses when you're doing exploratory analysis.

Of course, sometimes there is still causation – we just need to be very careful about inferring it. A somewhat touchy example of this is smoking and lung cancer. All we really know is that smoking is correlated with getting cancer down the road. From a purely statistical perspective, it's possible that some people have an underlying lung condition that makes them susceptible to cancer and also makes them prone to nicotine addiction. In this case, we must leverage our knowledge of biology and medicine, which gives compelling mechanisms for how a causal relationship would work. It's not rigorous statistical certainty, but we can still reach a scientific conclusion. This is one of my favorite examples of how rigorous math can dovetail with common sense and domain expertise.

5.13 Anscombe's Quartet and the Limits of Numbers

I've mentioned a number of times the limits of summary metrics and the fact that important features of a dataset can be masked by them. There is a famous madeup dataset called Anscombe's quartet that illustrates this fact.

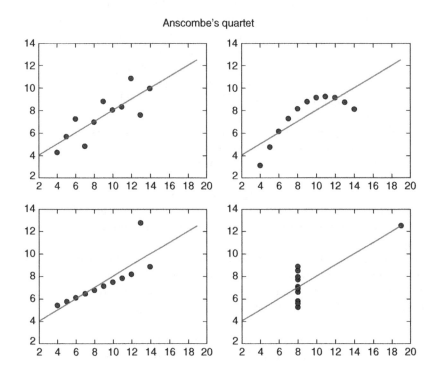

Anscombe's quartet

The following plot shows Anscombe's four datasets as scatterplots, along with their lines of best fit. In each of the four datasets, x and y have exactly the same average and standard deviation. Furthermore, there is the same correlation *between* x and y, and the lines of best fit are identical:

The first two plots show very different relationship between x and y: linear but noisy, and nonlinear by very clean. The other plots show the massive effect that outliers can have, either to throw off your best-fit parameters or to suggest that there is a "fit" when there really isn't one.

5.14   Time Series

Time series data is one of the most important data types in the world, capturing everything from stock prices, to website traffic, to blood glucose levels. As sensors become more ubiquitous, this importance will only grow. So, it's perhaps surprising that time series analysis is one of things that data scientists tend to be bad at: there is a rich set of time series techniques that are common practice in engineering, but that (largely for historical reasons, I think) just haven't really percolated into the data science community. I'll talk more about the analysis of time series data in a later chapter, but for now I just want to go over a few points about visualization techniques.

The first and most important thing is just how critical visualization is. Anscombe's quartet shows that relying on summary statistics can be dangerous, but reasonable bell curves are common enough in the real world that you can often get away with it. With time series though, there is absolutely no substitute for plotting. The pertinent pattern might end up being a sharp spike followed by a gentle taper down. Or, maybe there are weird plateaus. There could be noisy spikes that have to be filtered out. A good way to look at it is this: means and standard deviations are based on the naïve assumption that data follows pretty bell curves, but there is no corresponding "default" assumption for time series data (at least, not one that works well with any frequency), so you always have to look at the data to get a sense of what's normal.

A simple example of a time series plot in Python is here. Unfortunately, there isn't a good example of time series data built-in to scikit-learn, so I'm pulling one in from another library called statsmodels, which describes measurements of atmospheric CO_2 levels over many years.

```
import statsmodels.api as sm
dta = sm.datasets.co2.load_pandas().data
dta.plot()
plt.title("CO2 Levels")
```

```
plt.ylabel("Parts per million")
plt.show()
```

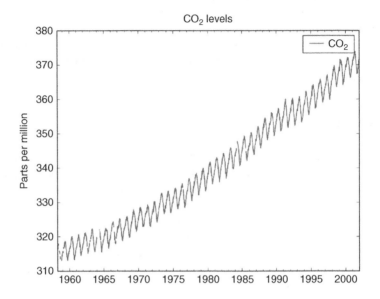

In this dataset, the DataFrame's index is set to a datetime type. In that case, Pandas is smart enough to do some very user-friendly formatting on the x-axis.

We can see in the plot that CO_2 fluctuates on a yearly cycle and that it increases overall across time. But there's no way we would have known to expect that a priori, unless we are already familiar with the science of CO_2. Image data is the same way; there is a huge amount of information and no a priori knowledge of how to extract features out of it, but the patterns are glaringly obvious to the human eye.

Along the lines of figuring out what patterns to expect, when you are exploring time series data, it is immensely useful to be able to zoom in and out. I have often zoomed in on a sharp spike only to find that it's actually a short-lived plateau. Or, what looked like an immediate step down turned into an exponential decay when I looked closer.

In some cases, you might want to plot not just the data itself but the log of the data (or, alternatively, plot it on a logarithmic scale). This makes sense with something such as the behavior of the price of a stock over a long period of time: a 10% uptick in the price will look equally impressive – which is good, since it's equally relevant to investment returns – whether the starting price was $1 or $20.

A good example is this code, which plots the price of Google's stock since 2000 on normal and logarithmic axes:

```
import urllib
import matplotlib.pyplot as plt
import pandas as pd
import numpy as np
# Get raw CSV data from the web
URL = ("http://ichart.finance.yahoo.com/" +
    "table.csv?s=GOOG&c=2000")
dat = urllib.urlopen(URL).read()
open('foo.csv','w').write(dat)
# Make DataFrame, w timestamp as the index
df = pd.read_csv('foo.csv')
df.index = df['Date'].astype('datetime64')
df['LogClose'] = np.log(df['Close'])
df['Close'].plot()
plt.title("Normal Axis")
plt.show()
df['Close'].plot(logy=True)
plt.title("Logarithmic Axis")
plt.show()
```

From the normal plot, it looks like Google had a massive surge in both 2005 and 2013. Indeed it did, in absolute dollar terms. But the logarithmic plot makes it clear that the surge in 2005 was much more significant, because it was a much larger increase proportionally.

Another common situation is that you have many different time series and you are looking for some kind of shared pattern. Plotting them all on the same chart (possibly after normalizing them, so they're all on a similar scale) is a useful way to do this. If there are so many that this becomes cluttered, what you can do is plot the median and quantiles; this makes it hard to see what is normal for an individual series but gives a good sense of how they move as a whole. Alternatively, you can plot the quantiles and overlay them with the plots of a manageably small sample of real time series.

Also, in many applications, it makes sense to look not just at the time series data itself but at transformations of it into the frequency domain. If you're measuring heart rate, temperature across days, or anything else where there is a reasonable expectation of periodicity, it can be immensely helpful to take a Fourier transform (see the later chapter if you're not familiar with these) and break the signal down into its component frequencies.

Also, you should still keep tools such as histograms in mind. Especially in data that is so noisy, it's hard to make sense of visually, and a histogram of the

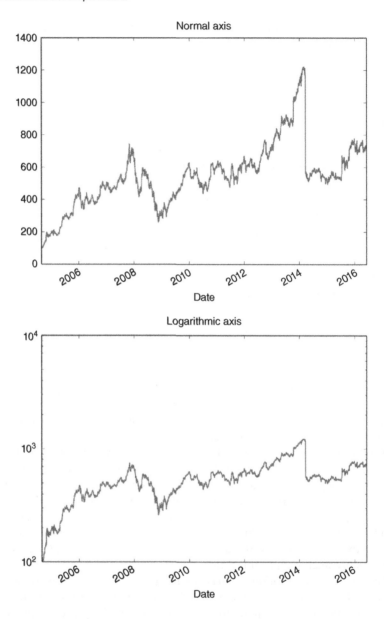

values can give you a useful summary of a time period. Bear in mind that for many applications, the visualizations are only a means to an end, an inspiration for how we can extract meaningful features to plug into a machine learning model. Some such as the median value over the time period are very reasonable features.

5.15   Further Reading

1 Janert, P, *Data Analysis with Open Source Tools*, 2010, O'Reilly Media, Newton, MA.
2 *Pandas: Python Data Analysis Library*, viewed 7 August 2016, http://pandas.pydata.org/.
3 *Matplotlib 1.5.1 Documentation*, viewed 7 August 2016, http://matplotlib.org/.

5.16   Glossary

Kendall correlation An ordinal correlation metric that is reasonably robust to gross outliers but highly sensitive to tiny variations in rank.

Logarithmic plot A plot where one or both axes are scaled to the logarithm of the value they portray. This makes it easier to visualize very small and very large numbers on a single axis.

Nonparametric correlation A correlation metric that doesn't tacitly assume a specific form for the relationship between two variables. Kendall and Spearman correlations are examples, since they only assume that the relationship is monotonic.

Pearson correlation The usual definition of correlation. Technically, $\text{Corr}[X,Y] = \text{Cov}[X,Y]/(\text{Std}[X] * \text{Std}[Y])$.

Quantile The Xth quantile is the value v such that a fraction X of your data points are $\leq v$, and a fraction $(1 - X)$ are equal to v.

Spearman correlation An ordinal correlation metric that is robust to small changes in ordering but can get thrown off badly by outliers.

6

Machine Learning Overview

In my mind, machine learning is technically a subset of statistics. However, that's not how it might look from the outside. For historical reasons, machine learning has evolved largely independently from statistics, in some cases reinventing the same techniques and giving them a different name, and in other cases inventing whole new ideas without statisticians supposedly involved. Classical statistics grew largely out of the needs of governments in processing census data and the agriculture industry. Machine learning evolved later and largely as an outgrowth of computer science. Early computer scientists, in turn, were drawn from the ranks of physicists and engineers. So the DNA is quite different, and the tools have diverged a lot, but ultimately they're tackling the same problems.

"Machine learning" has become a catchall term that covers a lot of different areas, ranging from classification to clustering. As such, I can't really give you a crisp definition of what it means. However, there are several commonalities that pretty much all machine learning algorithms seem to work with:

- It's all done using computers, leveraging them to do calculations that would be intractable by hand.
- It takes data as input. If you are simulating a system based on some idealized model, then you aren't doing machine learning.
- The data points are thought of as being samples from some underlying "real-world" probability distribution.
- The data is tabular (or at least you can think of it that way). There is one row per data point and one column per feature. The features are all numerical, binary, or categorical.

The last of these properties is the real kicker. Most machine learning algorithms are designed to handle pretty much any tabular dataset, but they ONLY handle tabular data.

Tabular data lends itself to all kinds of mathematical analysis, since the rows of a table with n rows and d columns can be viewed as locations in *d*-dimensional space. This is why machine learning is easily the most mathematically sophisticated thing a data scientist is likely to do.

In most machine learning applications, the data points are thought of as being drawn from some underlying distribution, and the goal is to find patterns in the samples that tell us something about that distribution as a whole or that will let us process other samples from it.

The next chapter, and several chapters in the next book part, will discuss several of the main areas of machine learning. The purpose of this short chapter is to give you some background about machine learning as a whole and some of the techniques that touch all parts of it.

6.1 Historical Context

Machine learning was partly born out of the initial failures of the artificial intelligence (AI) movement. For a long time, people were very focused on the idea that computers could be made to think, and it was widely expected that thinking machines were only a few years away. There is an anecdote that Marvin Minsky, one of the founders of AI, once assigned a grad student the task of working out computer vision over the course of a summer. People were thinking about the human brain as a big logic engine, and a lot of the focus was on getting computers to mimic the logical processing that humans do.

AI failed (at least relative to the hype it had generated), and it's partly out of embarrassment on behalf of their discipline that the term "artificial intelligence" is rarely used in computer science circles (although it's coming back into favor, just without the over-hyping). We are as far away from mimicking human intelligence as we have ever been, partly because the human brain is fantastically more complicated than a mere logic engine.

The focus has shifted away from creating true intelligence and toward using computers to do tasks that historically a human has to do. This includes things such as recognizing whether there is a bird in a photograph, telling whether an e-mail is spam, or identifying that an "interesting event" has occurred in a time series. Machine learning was built up on using computers as proxies for human judgment in specific, limited situations. Of course, the techniques thus developed can be applied to many areas, even ones where human judgment is never applied in practice, so machine learning has matured into a standard toolset for any data scientist.

The kinds of tools being used shifted as well. AI traditionally took a rule-based approach that used logical inference to reach conclusions. Machine learning is much more probabilistic in the way it makes models and inferences.

As a final note, some people might criticize this book for not going into nearly enough depth, especially with regard to cutting-edge developments in areas such as deep learning. The reason for this is simple: in my experience, data scientists rarely get that far into the weeds. Machine learning experts spend a lot of time improving their classifiers with all the latest tricks, but data scientists tend to use off-the-shelf classifiers, instead pouring their effort into finding good features to plug into them.

6.2 Supervised versus Unsupervised

There are two main types of machine learning, called supervised and unsupervised.

In supervised learning, your training data consists of some points and a label or target value associated with them. The goal of the algorithms is to figure out some way to estimate that target value. For example, we might have data on several medical patients saying what was in a drop of their blood and then whether they were later found to have cancer. If we want to use blood samples from future patients to assess their cancer risk, this is a supervised learning problem.

In unsupervised learning, there is just raw data, without any particular thing that is supposed to be predicted. Unsupervised algorithms are used for finding patterns in the data in general, teasing apart its underlying structure. Clustering algorithms, which try to break a dataset down into "natural" clusters, are a prototypical example of unsupervised learning.

Supervised learning is somewhat more common in real applications. Business situations usually dictate a specific thing that you are trying to predict, rather than a broad "see what there is to see" approach. However, unsupervised learning algorithms are often used as a preprocessing step for extracting meaningful features from a data point, with those features ultimately getting used for supervised learning.

6.3 Training Data, Testing Data, and the Great Boogeyman of Overfitting

By far the greatest headache in machine learning is the problem of overfitting. This means that your results look great for the data you trained them on, but they don't generalize to other data in the future. As an extreme case, imagine that your dataset of medical patients included their names and your trained classifier just remembered the names of everybody who had cancer and made predictions based on that. It would give perfect predictions for everybody it was trained on but would be useless for assessing anybody else's cancer risk.

The solution is to train on some of your data and assess performance on other data. This can be done in a number of ways:

- Most basically, you randomly divide your data points between training and testing. Randomness is critically important, so as to avoid unintentional sources of bias (such as taking the first half of your data file as training data, when those rows might have been collected earlier). Honestly, this crudely simple approach is often good enough in practice.
- A fancier method that works specifically for supervised learning is called k-fold cross-validation. The goal here isn't to measure the performance of a particular, fitted classifier, but rather a family of classifiers. Cross-validation is done this way:
  - Divide the data randomly into k partitions.
  - Train a classifier on all but one partition, and test its performance on the partition that was left out.
  - Repeat, but choosing a different partition to leave out and test on. Continue for all the partitions, so that you have k different trained classifiers and k performance metrics for them.
  - Take the average of all the metrics. This is the best estimate of the "true" performance of this family of classifiers when it is trained on this kind of data.
- If you're being very rigorous about your statistics, it is common to divide your data into a training set, a testing set, and a *validation* set. You only get to examine the validation set at the very end to test your hypotheses and the performance of your model. This is done to avoid a very subtle form of statistical bias. Let's say you only had testing/training data, and you had several machine learning models to choose from. In this case, you would pick the one that performed best when you trained it on the one dataset and tested it on the other. But this is a weak form of training on the test data, because the test data influences your choice of model. The validation data then lets you put your trained model to the *real* test.
- I don't know a name for it, but there is another approach I've used that is great in many real applications. Oftentimes, there's a situation where a model is retrained periodically, say every week, incorporating the new data acquired in the previous week. In these cases, it makes sense to train on all the data for week N and the previous weeks and then test it on all the data for week $N+1$. The reasons why people click on ads might have changed a little bit over the course of that week, making the model slightly outdated, so testing on data from the same time period can artificially inflate your performance.

6.4   Further Reading

1 Bishop, C, *Pattern Recognition and Machine Learning*, 2007, Springer, New York, NY.
2 *Scikit-learn 0.171.1 documentation*, http://scikit-learn.org/stable/index.html, viewed 7 August 2016, The Python Software Foundation.

6.5   Glossary

Artificial intelligence Trying to mimic human-like reasoning and behavior in a computer program. Artificial intelligence fell from grace when the fields failed to live up to its hype. Machine learning solves a lot of similar problems, but it does so using statistical techniques rather than rule-based ones, and it usually makes no pretense of mimicking the human brain.

Machine learning A catchall term for several techniques that operate on tabular data.

Overfitting A machine learning model becoming so specially tuned to its exact input data that it fails to generalize to other, similar data.

Supervised learning Machine learning where there is a specific target variable you are trying to predict per data point.

Tabular data A dataset that is arranged in rows and numerical columns. Each row is associated with some entity, and each column gives some feature about all the entities.

Testing data Data that is used to assess how well a machine learning model performs. It should not have been involved in the creation of that model.

Training data Data that is used for training a machine learning model. The performance of the model should generally not be tested on the training data.

Unsupervised learning Machine learning where there is *not* a specific target variable you are trying to predict. Clustering is an example.

7

Interlude: Feature Extraction Ideas

Before we jump into specific machine learning technique, I want to come back to feature extraction. A machine learning analysis will be only as good as the features that you plug into it. The best features are the ones that carefully reflect the thing you are studying, so you're likely going to have to bring a lot of domain expertise to your problems. However, I can give some of the "usual suspects": classical ways to extract features from data that apply in a wide range of contexts and are at the very least worth taking a look at. This interlude will go over several of them and lead to some discussion about applying them in real contexts.

7.1   Standard Features

Here are several types of feature extraction that are real classics, along with some of the real-world considerations of using them:

- Is_null: One of the simplest, and surprisingly effective, features is just whether the original data entry is missing. This is often because the entry is null for an important reason. For example, maybe some data wasn't gathered for widgets produced by a particular factory. Or, with humans, maybe demographic data is missing because some demographic groups are less likely to report it.
- Dummy variables: A categorical variable is one that can take on a finite number of values. A column for a US state, for example, has 50 possible values. A dummy variable is a binary variable that says whether the categorical column is a particular value. Then you might have a binary column that says whether or not a state is Washington, another column that says whether it is Texas, and so on. This is also called one-hot encoding, because every row in your dataset will have 1 in exactly one of the dummy variables for the states. There are two big issues to consider when using dummy variables:
  a) You might have a LOT of categories, some of which are very rare. In this case, it's typical to pick some threshold and only have dummy variables

The Data Science Handbook, First Edition. Field Cady.
© 2017 John Wiley & Sons, Inc. Published 2017 by John Wiley & Sons, Inc.

for the more common values, then have another dummy variable that will be 1 for anything else.

b) Often, you only learn what the possible values are by looking at training data, and then you will have to extract the same dummy features from other data (maybe testing data) later on. In that case, you will have to have some protocol for dealing with entries that were not present in the training data.

- Ranks: A blunt-force way to correct for outliers in a column of data is to sort the values and instead use their ordinal ranks. There are two big problems with this:
  a) It's an expensive computation since the whole list must be sorted, and it cannot be done in parallel if your data is distributed across a cluster.
  b) Ranks are a huge problem when it comes to testing/training data. If you rank all your points before dividing into training/testing, then information about the testing data will be implicit in the training data: a huge no-no. A workaround is to give each testing data point the rank that it *would* have had in the training data, but this is computationally expensive.

- Binning: Both of the problems associated with ranks can be addressed by choosing several histogram bins of roughly equal size that your data can be put into. You might have a bin for anything below the 25th percentile, another for the 25th to the 50th percentile, and so on. Then rather than a percentile rank for your data points, just say which bin they fall into. The downside is that this takes away the fine resolution that you get with percentile ranks.

- Logarithms: It is common to take the logarithm of a raw number and use that as a feature. It dampens down large outliers and increases the prominence of small values. If your data contains any 0s, it's common to add 1 before taking the log.

7.2 Features That Involve Grouping

Oftentimes, a dataset will include multiple rows for a single entity that we are describing. For example, our dataset might have one row per transaction and a column that says the customer we had the transaction with, but we are trying to extract features about the customers. In these cases, we have to aggregate the various rows for a given customer in some way. Several brute-force aggregate metrics you use could include the following:

- The number of rows
- The average, min, max, mean, median, and so on, for a particular column

- If a column is nonnumerical, the number of distinct entries that it contains
- If a column is nonnumerical, the number of entries that were identical to the most common entry
- The correlation between two different columns.

7.3 Preview of More Sophisticated Features

Many of the more advanced chapters in this book will talk about fancy methods for feature extraction. Here is a quick list of some of the very interesting ones:

- If your data point is an image, you can extract some measure of the degree to which it resembles some other image. The classical way to do this is called Principal Component Analysis (PCA). It also works for numerical arrays of time series data or sensor measurements.
- You can cluster your data and use the cluster of each point as a categorical feature.
- If the data is text, you can extract the frequency of each word. The problem with this is that it often gives you prohibitively many features, and some additional method may be required to condense them down.

7.4 Defining the Feature You Want to Predict

Finally, it's worth noting that you may find yourself extracting the most important feature of all: the feature that you're using machine learning to try and predict. To show you how this might work, here are several examples from my own career:

- I had to predict human traffic to a collection of websites. However, the traffic logs were polluted by bots, and it became a separate problem to filter out the bot traffic and then estimate how well our cleaned traffic corresponded to flesh-and-blood humans. We did this by comparing our traffic estimate to those from Google for a few select sites – sometimes we were over, and sometimes we were under, but we concluded that it was a good enough match to move forward with the project.
- I have studied customer "churn," that is, when customers take their business elsewhere. Intuitively, the "ground truth" is a feeling of loyalty that exists in the minds of customers, and you have to figure out how to gauge that based on their purchasing behavior. It's hard to distinguish between churn and the customer just not needing your services for a time period.

- When you are trying to predict events based on time series data, you often have to predict whether or not an event is imminent at some point in time. This requires deciding how far in the future an event can be before it is considered "imminent." Alternatively, you can have a continuous-valued number that says how long until the next event, possibly having it top out at some maximum value so as to avoid outliers (or you could take the logarithm of the time until the next event – that would dampen outliers too).

8

Machine Learning Classification

Machine learning classifiers are a critically important part of the data science toolkit. However, they are not nearly as important as they are made out to be. A large part of the mystique of data science comes from the idea that we can pour data into a magical black box that (through some mathematical voodoo that only data scientists are smart enough to understand) can learn everything about the data and solve business problems.

The reality is a lot more mundane. As we've discussed previously, it takes a lot of work to get the data into a form where it can be fed into the black box, a lot of savvy to point the black box at the right question, and additional work to make sense of the results. The machine learning black box itself is usually just a library that you call. Sure, it's good to have some idea of how the classifiers work under the hood – you can pick better ones to use, avoid common pitfalls, make better sense of their output, and understand how to jury-rig them as need be. But training a plain-vanilla classifier is often construed as being rocket science, and it's not.

This chapter comes in two sections. After some initial notes, the first will be a series of rapid-fire tutorials about some of the most useful classifiers. The second section will discuss the various ways that we can grade their accuracy.

8.1　What Is a Classifier, and What Can You Do with It?

A machine learning classifier is a computational object that has two stages:

- It gets "trained." It takes in its training data, which is a bunch of data points and the correct label associated with them, and tries to learn some pattern for how the points map to the labels.
- Once it has been trained, the classifier acts as a function that takes in additional data points and outputs predicted classifications for them. Sometimes,

The Data Science Handbook, First Edition. Field Cady.
© 2017 John Wiley & Sons, Inc. Published 2017 by John Wiley & Sons, Inc.

the prediction will be a specific label; other times, it will give a continuous-valued number that can be seen as a confidence score for a particular label.

There are two big use cases for classifiers. The first is the obvious one; we have things that need to be classified. This happens all the time in production code situations, where a computer has to decide, say, which ad to show a user. It also happens when computers aren't making decisions autonomously, but instead flagging things for a flesh-and-blood human to look at: flagging potential instances of credit card fraud, for example.

The other use for classifiers is to give insights about the underlying data. In my own career, this has actually been the more common use case. My clients have not been so interested in predicting, say, that a particular machine will fail. What they really want to know is the patterns in the data that predict failures, because those patterns can help them diagnose and fix something that's going wrong on their assembly line. In cases such as this, we want to dissect our classifiers after the fact, extracting business insights. This becomes an interesting balancing act for a data scientist; sometimes, the most accurate classifiers are the hardest to make real-world sense of.

8.2 A Few Practical Concerns

The whole notion of machine learning classification is premised on the idea of having correctly labeled training data and in sufficient quantities to train our classifier. However, this is a luxury that the real world often doesn't afford. For example, in fraud detection, you will probably have a modest-size set of hand-labeled fraud cases and a huge mass of unlabeled data. You just presume that those unlabeled points are nonfraudulent, which means that an unknown fraction of your training data is mislabeled. It's not like the hand-labeled fraud cases are a nice random sample from all fraud cases; they represent whatever kind of fraud people have been looking for so far. There could easily be whole new categories of fraud, every instance of which is labeled as nonfraud in your training data.

What even is "fraud" anyway? In many cases, you have to come up with the training data yourself, and it's not a priori clear how things should be labeled. An e-mail about Nigerian princes is almost certainly fraud, but what about somebody selling "discont Vi@gra?"

In cases where I'm looking for cool patterns, rather than the classifiers themselves, what I will often do is remove edge cases from my training data. For example, I once had a client who was trying to understand "customer loyalty," and I was writing a classifier for whether or not they would lose a customer within the next year. The problem is that customers don't announce they're leaving – they just stop using my client's services, which most customers don't use all that often anyway. What I did was to formulate criteria for customers being "definitely still loyal," and another for them being "definitely not loyal." Every "gray area" customer, who didn't fall into one of those categories, was discarded before training – they made up around a third of customers. The resulting classifier worked surprisingly well. But the really exciting part was that when we applied it to the gray area customers, those that were flagged as higher-risk did indeed come closer to satisfying the "not loyal" criteria compared to the loyal criteria. This suggested that (not surprisingly, but reassuring to know) the gray area customers were more- or less-extreme versions of what the other customers were doing, rather than some fundamentally new category of people.

This question of defining ground truth is beside the perennial problem of data science: feature extraction. Contrary to popular myth, any machine learning classifier will suck if you give it features that don't contain signal or features for which the signal is deeply buried in idiosyncrasies and convoluted dependencies. A large portion of data science boils down to understanding the dataset and the domain of application well enough that you can extract meaningful features.

8.3 Binary versus Multiclass

Most classification problems have a binary classification: 1 or 0, yes or no. However, oftentimes, the label is a categorical variable, capable of taking on several values. There are some classifier algorithms that handle this situation natively, but many others are strictly binary. When you are using a classifier that is binary but solving a problem that has k possibly labels, the standard solution is to actually train k different classifiers: one for each label X, classifying points as being X or something else.

For the most part, these distinctions are wrapped up within machine learning libraries and invisible to data scientists who use the libraries. Explanations in this chapter will freely assume that classifiers are all binary.

8.4 Example Script

The following script demonstrates many of the topics we will cover in this chapter in a realistic setting, using the sample iris dataset from the last chapter. It takes several important classifiers, trains each one to distinguish iris virginica from the other species, and then plots the results on an ROC curve (I'll explain these shortly – they're a tool for visualizing how well a classifier works). Each of these classifiers, and the metrics we use to evaluate them, will be explained later in the chapter.

```python
from matplotlib import pyplot as plt
import sklearn
from sklearn.metrics import roc_curve, auc
from sklearn.cross_validation import train_test_split
from sklearn.linear_model import LogisticRegression
from sklearn.tree import DecisionTreeClassifier
from sklearn.ensemble import RandomForestClassifier
from sklearn.naive_bayes import GaussianNB

# name -> (line format, classifier)
CLASS_MAP = {
    'LogisticRegression':
        ('-', LogisticRegression()),
    'Naive Bayes': ('--', GaussianNB()),
    'Decision Tree':
        ('.-', DecisionTreeClassifier(max_depth=5)),
    'Random Forest':
        (':', RandomForestClassifier(
            max_depth=5, n_estimators=10,
            max_features=1)),
    }

# Divide cols by independent/dependent, rows by test/
train
X, Y = df[df.columns[:3]], (df['species']=='virginica')
X_train, X_test, Y_train, Y_test = \
    train_test_split(X, Y, test_size=.8)

for name, (line_fmt, model) in CLASS_MAP.items():
    model.fit(X_train, Y_train)
    # array w one col per label
    preds = model.predict_proba(X_test)
    pred = pd.Series(preds[:,1])
    fpr, tpr, thresholds = roc_curve(Y_test, pred)
    auc_score = auc(fpr, tpr)
    label='%s: auc=%f' % (name, auc_score)
    plt.plot(fpr, tpr, line_fmt,
        linewidth=5, label=label)

plt.legend(loc="lower right")
plt.title('Comparing Classifiers')
```

```
plt.plot([0, 1], [0, 1], 'k--') #x=y line.  Visual aid
plt.xlim([0.0, 1.0])
plt.ylim([0.0, 1.05])
plt.xlabel('False Positive Rate')
plt.ylabel('True Positive Rate')
plt.show()
```

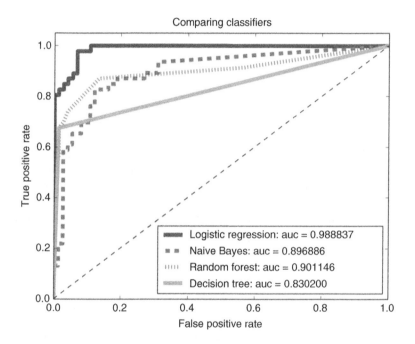

8.5 Specific Classifiers

The world is full of different classification algorithms. This section will go over some of the most useful and important ones.

8.5.1 Decision Trees

A decision tree is conceptually one of the simplest classifiers available. Using a decision tree to classify a data point is the equivalent of following a basic flow-chart. It consists of a tree structure such as the following:

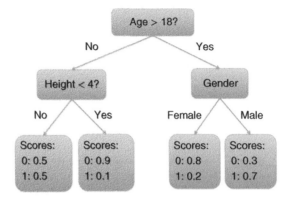

Every node in the tree asks a question about one feature of a data point. If the feature is numerical, the node asks whether it is above or below a threshold, and there are child nodes for "yes" and "no." If the feature is categorical, typically there will be a different child node for every value it can take. A leaf node in the tree will be the score that is assigned to the point being classified (or several scores, one for each possible thing the point could be flagged as). It doesn't get much simpler than that.

Using a decision tree is conceptually quite straightforward, but training one is another matter entirely. In general, finding the optimal decision for your training data is computationally intractable, so in practice, you train the tree with a series of heuristics and hope that the result is close to the best tree possible. Generally, the algorithm is something along these lines:

1) Given your training data X, find the single feature (and cutoff for that feature, if it's numerical) that best partitions your data into classes.
2) There are a variety of ways to quantify how good a partition is. The most common ones are the "information gain" and the "Gini impurity." I won't delve into their precise meanings here.
3) This single best feature/cutoff becomes the root of your decision tree. Partition X up according to this node.
4) Recursively train each of the child nodes on its partition of the data.
5) The recursion stops when either all of the data points in your partition have the same label or the recursion has gone to a predetermined maximum depth. At that point, the scores stored in this node will just be the breakdowns of the labels in the partition.

You will probably never need to worry about the details of how to train a decision tree, but knowing the basic process helps to understand one of the biggest problems with this classifier: overfitting. If you set the maximum depth too far, for example, every leaf node will end up with a partition that contains only a few points, all of which have the same label. This will result in the decision tree consistently giving out extremely confident scores that, realistically, are just an accident of small numbers. You can set parameters on your decision tree that force it to terminate when, say, the best partitions on a node are too small. But in many libraries, the default settings will let a decision tree drastically overfit itself, so you should be aware of this and tune your trees accordingly.

Decision trees are very easy to understand, so it's perhaps a bit surprising that they are often difficult to tease real-world insights out of. Looking at the top few layers is certainly interesting and suggests what some of the more important features are. But it's not always clear what the features and their cutoffs mean to the real world, unless you want to wade into the deep waters of dissecting the Gini impurities from the training stage. Even if you do this, there is still a very real risk that the same feature will weigh toward hits at one node of the tree and toward nonhits at another node. What the heck does that mean?

Personally, I don't use decision trees much for serious work. However, they are extremely useful for their human readability – this is especially handy if you're working with people who don't know machine learning and are wary of black boxes – and the rapidity with which they can do classifications. Above all, decision trees are useful as a building block for constructing Random Forest classifiers, which I'll discuss in the next section.

The following code shows how to train and use a decision tree in Python:

```
from sklearn.tree import DecisionTreeClassifier
clf = DecisionTreeClassifier(max_depth=5)
clf.fit(train[indep_cols], train.breed)
predictions = clf.predict(test[indep_cols])
```

8.5.2 Random Forests

If I were stuck on a desert island and could only take one classifier with me, it would be the random forest. They are consistently one of the most accurate, robust classifiers out there, legendary for taking in datasets with a dizzying number of features, none of which are very informative and none of which have been cleaned, and somehow churning out results that beat the pants off of anything else.

The basic idea is almost too simple. A random forest is a collection of decision trees, each of which is trained on a random subset of the training data and only allowed to use some random subset of the features. There is no coordination in the randomization – a particular data point or feature could randomly get plugged into all the trees, none of the trees, or anything in between. The final classification score for a point is the average of the scores from all the trees (or sometimes, you treat the decision trees as binary classifiers and report the fraction of all of them that votes a certain way).

The hope is that the different trees will pick up on different salient patterns, and each one will only give confident guesses when its pattern is present. That way, when it comes time to classify a point, several of the trees will classify it correctly and strongly while the other trees give answers that are on the fence, meaning that the overall classifier slouches toward the right answer.

The individual trees in a random forest are subject to overfitting, but they tend to be randomly overfitted in different ways. These largely cancel each other out, yielding a robust classifier.

The problem with random forests is that they're impossible to make real business sense of. The whole point of a classifier such as this is that it is too complex for human comprehension, and its performance is an averaged-out thing.

The one thing that you can do with a random forest is to get a "feature importance" score for any feature in the dataset. These scores are opaque and impossible to ascribe a specific real-world meaning to. The importance of the kth feature is calculated by randomly swapping the kth feature around between the points in the training data and then looking at how much randomizing this feature hurts performance (there is a little bit of extra logic to make sure that no randomized data point is fed into a tree that was trained on its nonrandomized version). In practice, you can often take this list of features and, with a little bit of old-fashioned data analysis, figure out compelling real-world interpretations of what they mean. But the random forest itself tells you nothing.

The following code shows how to train and use a random forest in Python:

```
from sklearn.tree import RandomForestClassifier
clf = RandomForestClassifier(
            max_depth=5, n_estimators=10,
            max_features=1))
clf.fit(train[indep_cols], train.breed)
predictions = clf.predict(test[indep_cols])
```

8.5.3 Ensemble Classifiers

Random forests are the best-known example of what are called "ensemble classifiers," where a wide range of classifiers (decision trees, in this case) are trained under randomly different conditions (in our case, random selections of data points and features) and their results are aggregated. Intuitively, the idea is that if every classifier is at least marginally good, and the different classifiers are not very correlated with each other, then the ensemble as a whole will very reliably slouch toward the right classification. Basically, it's using raw computational power in lieu of domain knowledge or mathematical sophistication, relying on the power of the law of large numbers.

8.5.4 Support Vector Machines

I'll be honest: I personally hate support vector machines (SVMs). They're one of the most famous machine learning classifiers out there, so it's important that you be familiar with them; but I have several gripes. First off, they make a very strong assumption about the data called linear separability. Oftentimes that assumption is wrong, and occasionally it's right in mathematically perverse ways. There are sometimes hacks that work around this assumption, but there's no principle behind them and no a priori way of knowing which (if any) hack will work in a particular situation. SVMs are also one of the few classifiers that are fundamentally binary; they don't give continuous-valued "scores" that can be used to assess how confident the classifier is. This makes them annoying if you're looking for business insights and unusable if you need to have the notion of a "gray area."

That said, they're popular for a reason. They are intuitively simple, mathematically elegant, and trivial to use. Plus, those unprincipled hacks I mentioned earlier can be incredibly powerful if you pick the right one.

The key idea of an SVM is illustrated in the following figure:

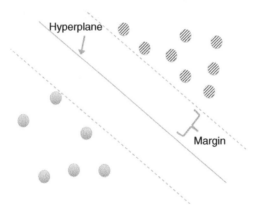

Essentially, you view every data point as a point in d-dimensional space and then look for a hyperplane that separates the two classes. The assumption that there actually is such a hyperplane is called linear separability. Training the SVM involves finding the hyperplane that (1) separates the datasets and (2) is "in the middle" of the gap between the two classes. Specifically, the "margin" of a hyperplane is min(its distance to the nearest point in class A, its distance to the nearest point in class B), and you pick the hyperplane that maximizes the margin.

Mathematically, the hyperplane is specified by the equation

$$f(x) = w \cdot x + b = 0$$

where w is a vector perpendicular to the hyperplane and b measures how far offset it is from the origin. To classify a point x, simply calculate $f(x)$ and see whether it is positive or negative. Training the classifier consists of finding the w and b that separates the dataset while having the largest margin.

This version is called "hard margin" SVM. However, in practice, there often is no hyperplane that completely separates the two classes in the training data. Intuitively, what we want to do is find the best hyperplane that *almost* separates the data, by penalizing any points that are on the wrong side of hyperplane. This is done using "soft margin" SVM.

The other killer problem with SVM is if you have as many features as data points. In this case, there is guaranteed to be a separating hyperplane, regardless of how the points are labeled. This is one of the curses of working in high-dimensional space. You can do dimensionality reduction (which I will discuss in a later chapter) as a preprocessing step, but if you just plug high-dimensional data into an SVM, it is almost guaranteed to be grotesquely overfitted.

The most notorious problem with a plain SVM is the linear separability assumption. An SVM will fail utterly on a dataset as the following:

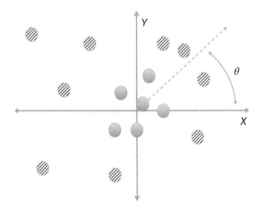

There is no line between the two classes of points. The pattern is clear if you just look at it – one class is near the origin, and the other is far from it – but an SVM can't tell. The solution to this problem is a very powerful generalization of SVM called "kernel SVM." The idea of kernel SVM is to first map our points into some other space in which the decision boundary is linear, and then construct a support vector machine that operates in that space. For the previous figure, if we plot the distance from the origin on the x-axis and the angle θ on the y-axis, we get the following figure:

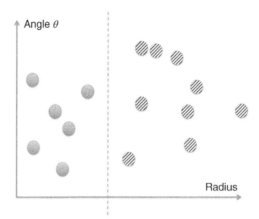

The data here is linearly separable. In general, kernel SVM requires finding some function ϕ that maps our points in d-dimensional space to points in some n-dimensional space. In the example I gave, n and d were both 2, but in practice, we usually want n to be larger than d, to increase the chances of linear separability. If you can find ϕ, then you're golden.

Now here's the key point computationally: you never need to find ϕ itself. When you crank through the math, it turns out that whenever you calculate $\phi(x)$, it is always part of a larger expression called the kernel function:

$$k(x,y) = \phi(x) \cdot \phi(y)$$

The kernel function takes two points in the original space and gives their dot product in the mapped space. This means that the mapping function ϕ is just an abstraction – we never need to calculate it directly and can instead just focus on k. In many cases, it turns out that calculating k directly is much, much easier than calculating any $\phi(x)$ intermediates. It is often the case that ϕ is an intricate mapping into a massively high-dimensional space, or even an infinite-dimensional space, but the expression for k reduces to some simple, tractable function that is nonlinear. Using only the kernel function in this way is called the "kernel trick," and it ends up applying to areas outside of SVMs.

Not every function that takes in two vectors is a valid kernel, but an awful lot of them are. Some of the most popular, which are typically built into libraries, are as follows:

- Polynomial kernel: $k(x,y) = (x \cdot y + c)^n$
- Gaussian kernel: $k(x,y) = \exp[-\gamma \, | \, x - y \, |^2]$
- Sigmoid: $k(x,y) = \tanh(x \cdot y + r)$.

Most kernel SVM frameworks will let users define their own functions as well. If you take this route, you should be aware that it's a bit technical to make sure that k is a valid kernel function, that is, that it has a corresponding mapping ϕ. Most simply k has to be symmetric: $k(x,y) = k(y,x)$ for any x and y. The major constraint though is that it be "positive definite." This is a highly technical constraint that I won't get into here.

8.5.5 Logistic Regression

Logistic regression is a great general-purpose classifier, striking an excellent balance between accurate classifications and real-world interpretability. I think of it as kind of a nonbinary version of SVM, one that scores points with probabilities based on how far they are from the hyperplane, rather than using that hyperplane as a definitive cutoff. If the training data is almost linearly separable, then all points that aren't near the hyperplane will get a confident prediction near 0 or 1. But if the two classes bleed over the hyperplane a lot, the predictions will be more muted, and only points far from the hyperplane will get confident scores.

In logistic regression, the score for a point will be

$$p(x) = \frac{1}{1 + \exp[w \cdot x + b]}$$

Note that $\exp[w \cdot x + b]$ is the same $f(x)$ we saw in SVM, where w is a vector that gives weights to each feature and b is a real-valued offset. With SVM we look at whether $f(x)$ is positive or negative, but in this case we plug it into the so-called "sigmoid function":

$$\sigma(z) = \frac{1}{1 + \exp[z]}$$

As with SVM, we have a dividing hyperplane defined by $w \cdot x + b = 0$. In SVM, that hyperplane is the binary decision boundary, but in this case, it is the hyperplane along which $p(x) = \frac{1}{2}$.

The sigmoid function shows up a few places in machine learning, so it makes sense to dwell on it a bit. If you plot out $\sigma(x)$, it appears as follows:

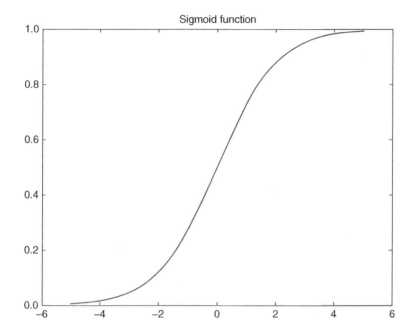

You can see that $\sigma(0)$ is 0.5. As the argument blows up to infinity, it approaches 1.0, and as it goes to negative infinity, it goes 0.0. Intuitively, this makes it a great way to take "confidence weights" and cram them down into the interval (0, 1.0) where they can be treated as probabilities. The sigmoid function also has a lot of convenient mathematical properties that make it easy to work with. We will see it again in the section on neural networks.

Pulling real-world meaning out of a trained logistic regression model is easy:

- If the kth component of w is large and positive, then the kth feature being big suggests that the correct label is 1.
- If the kth component of w is large and negative, then the kth feature being big suggests that the correct label is 0.
- The larger the elements of w are in general, the tighter our decision boundary and the more closely we approach an SVM.

Note though that in order for this to be meaningful, you must make sure that your data is all set to the same scale before training; if the most important feature also happens to be the largest number, then its coefficient would be misleadingly small.

Another perk of logistic regression is that it's extremely efficient to store and use. The entire model consists of just $d+1$ floating point numbers, for the d components of the weight vector and the offset b. Performing a classification requires just d multiplication operations, d addition operations, and one computation of a sigmoid function.

This code will train and use a logistic regression model:

```
from sklearn import linear_model
clf = linear_model.LogisticRegression()
clf.fit(train_data, train_labels)
predictions = clf.predict(test_data)
```

8.5.6 Lasso Regression

Lasso regression is a variant of logistic regression. One of the problems with logistic regression is that you can have many different features all with modest weights, instead of a few clearly meaningful features with large weights. This makes it harder to extract real-world meaning from the model. It is also an insidious form of overfitting, which is begging to have the model generalize poorly.

In lasso regression, $p(x)$ has the same functional form of $\sigma(w \cdot x + b)$. However, we train it in a way that punishes modest-sized weights. The numerical algorithm that finds the optimal weights generally doesn't use heuristics or anything; it's just cold numerical trudging. However, as an aid to human intuition, I like to think of some examples of heuristics that the solver might, in effect, employ:

- If features i and j have large weights, but they usually cancel each other out when classifying a point, set both their weights to 0.
- If features i and j are highly correlated, you can reduce the weight for one while increasing the weight for the other and keeping predictions more or less the same.

The end result of all this tends to be having most of the feature weights go to 0 while only a few of the most significant features have nonzero weights.

8.5.7 Naive Bayes

Bayesian statistics is one of the biggest, most interesting, and most mathematically sophisticated areas of machine learning. However, most of that is in the context of Bayesian networks, which are a deep, highly sophisticated family of models that you typically don't see in normal data science (although I will discuss them a little bit in a later chapter). Data scientists are more likely to use a drastically simplified version called naive Bayes.

I talk in more detail about Bayesian statistics in the chapter on statistics. Briefly though, a Bayesian classifier operates on the following intuition: you start off with some initial confidence in the labels 0 and 1 (assume that it's a binary classification problem). When new information becomes available, you adjust your confidence levels depending on how likely that information is conditioned on each label. When you've gone through all available information, your final confidence levels are the probabilities of the labels 0 and 1.

Ok, now let's get more technical. During the training phase, a naive Bayesian classifier learns two things from the training data:

- How common every label is in the whole training data
- For every feature X_i, its probability distribution *when the label is 0*
- For every feature X_i, its probability distribution *when the label is 1*.

The last two are called the conditional probabilities, and they are written as

$$\Pr(X_i = x_i \mid Y = 0)$$
$$\Pr(X_i = x_i \mid Y = 1)$$

When it comes time to classify a point $X = (X_1, X_2, ..., X_d)$, the classifier starts off with confidences

$$\Pr(Y = 0) = \text{fraction of the training data with Y=0}$$
$$\Pr(Y = 0) = \text{fraction of the training data with Y=1}$$

Then for each feature X_i in the data, let x_i be the value it actually had. We then update our confidences to

$$\Pr(Y = 0) \leftarrow \Pr(Y = 0) * \Pr(X_i = x_i \mid Y = 0) * \gamma$$
$$\Pr(Y = 1) \leftarrow \Pr(Y = 1) * \Pr(X_i = x_i \mid Y = 1) * \gamma$$

where we set γ so that the confidences add up to 1.

There are a lot of things here that need fleshing out if you're implementing a naive Bayes classifier. For example, we need to assume some functional form for $\Pr(X_i = x_i \mid Y = 0)$, such as a normal distribution or something, in order to fit it during the training stage. We also need to be equipped to deal with overfitting there.

But the biggest problem is that we are treating X_i as being independent of every other X_j. For example, our data might be such that X_5 is just a copy of X_4. In that case, we really shouldn't adjust our confidences when we get to X_5, since X_4 had already accounted for it. Naive Bayes classifiers completely ignore this possibility, so it's perhaps surprising that they tend to be very powerful classifiers. The way I think of it is this: imagine the situation I described, where X_4 and X_5 are identical, so we essentially double-count X_4. If X_4 is a powerful predictive variable, then this might make us overconfident, but it usually doesn't make us wrong.

The following code uses scikit-learn to train and use a Gaussian naive Bayes classifier, where $\Pr(x_{i/\gamma})$ is assumed to be a normal distribution:

```
from sklearn.naive_bayes import GaussianNB
clf = GaussianNB()
```

```
clf.fit(train[indep_cols], train.breed)
predictions = clf.predict(test[indep_cols])
```

8.5.8 Neural Nets

Neural nets used to be the black sheep of classifiers, but they have enjoyed a renaissance in recent years, especially the sophisticated variants collectively known as "deep learning." Neural nets are as massive an area as Bayesian networks, and many people make careers out of them. However, basic neural nets are standard tools that you see in machine learning. They are simple to use, fairly effective as classifiers, and useful for teasing interesting features out of a dataset.

Neural nets were inspired by the workings of the human brain, but now that we know more about how biological circuits work, it's clear that that analogy is bunk. Really sophisticated deep learning is at the point where it can be compared to some parts of real brains (or maybe we just don't know enough about the brain yet to see how much they fall short), but anything short of that should be thought of as just another classifier.

The simplest neural network is the perceptron. A perceptron is a network of "neurons," each of which takes in multiple inputs and produces a single output. An example is shown in the following figure:

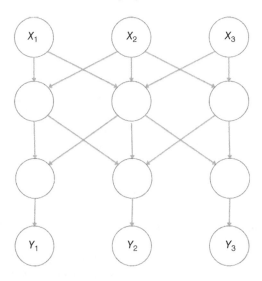

The labeled nodes correspond to either the input variables to a classification or a range of output variables. The other nodes are neurons. The neurons in the first layer take all of the raw features as inputs. Their outputs are fed as inputs to the second layer, and so on. Ultimately, the outputs of the final layer constitute the output of your program. All layers of neurons before the last one are called "hidden" layers (there is one in this figure the way I've drawn it). Unlike other classifiers, neural networks very organically produce an arbitrary number of different outputs, one for each neuron in the final layer. In this case, there are three outputs. In general, you can use neural nets for tasks other than classification and treat the outputs as a general-purpose numerical vector. In classification tasks though, we typically look at the ith output as the score for the ith category that a point can be classified as.

The key part of a neural net is how each neuron determines its output from its various inputs. This is called the "activation function," and there are a number of options you can pick from. The one I've seen the most is our old friend, the sigmoid function. If we let i indicate some particular neuron, and j ranges over its inputs, then:

$$\text{Activation}_i = \sigma\left(b_i + \sum_j w_{ij} * \text{Input}_j \right)$$

In effect, each neuron in the system is its own little logistic regression function, operating on the inputs to that neuron. A neural network with no hidden layers is, in fact, just a collection of logistic regressors.

Neural networks are trained using an iterative algorithm called backpropagation. Without getting into too much detail, you feed it sequences of input vectors along with the corresponding correct output vectors (in classification, each correct output vector would be all 0s except for a single one). As you do this, the parameters in the final layer get tweaked to accommodate the new information. Those changes are sent back to the previous layer, whose parameters are tweaked, and so on. Part of the beauty of the sigmoid function is that it makes backpropagation more mathematically tractable.

Many classifiers are useful for identifying single features that are especially informative. Neural networks, on the other hand, are known for identifying aggregate features, which are often more interesting than any of the raw inputs. Let's say you have a neural network with one hidden layer. A neuron in the hidden layer works by taking a weighted combination of all the raw inputs and plugging that combination in a sigmoid function. Well that weighted combination is an aggregate feature of your data and often an extremely useful one. If you take a bunch of images of handwritten letters and train a neural net on them, the input layer will be raw pixels and the 26 outputs will correspond to letters of the alphabet. But neurons in the middle will emit signals that correspond to things such as straight line segments, curves, and other key components of letters. Going back and making real-world sense of a neural network's internal features can be very illuminating.

For myself, neural networks are not a tool I use a lot. Simple ones such as the perceptron don't perform particularly well, and using the more complicated ones is an extremely specialized area. I'm more of an ensemble classifier guy, trusting in the law of large numbers rather than the voodoo of deep learning. But that's just me. Neural nets are a hot area, and they are solving some very impressive problems. There's a good chance they will become a much larger, more standard tool in the data science toolkit.

8.6 Evaluating Classifiers

In most business applications of classification, you are looking for one class more so than the other. For example, you are looking for promising stocks in a large pool of potential duds. Or, you are looking for patients who have cancer. The job of a classifier is to flag the interesting ones.

There are two aspects of how well a classifier performs: you want to flag the things you're looking for, but you also want to not flag the things you aren't looking for. Flag aggressively, and you'll get a lot of false positives – potentially very dangerous if you're looking for promising stocks to invest in. Flag conservatively, and you'll be leaving out many things that should have been flagged – terrible if you're screening people for cancer. How to strike the balance between false positives and false negatives is a business question that can't be answered analytically.

In this chapter, we will focus on two performance metrics that, together, give the full picture of how well a classifier performs:

- True positive rate (TPR). Of all things that should be flagged by our classifier, this is the fraction that actually gets flagged. We want it to be high: 1.0 is perfect.
- False positive rate (FPR). Of all things that should NOT be flagged, this is the fraction that still ends up getting flagged. We want it low: 0.0 is perfect.

I will give you a nice graphical way to think of TPR and FPR, that is, the main way I think about classifiers in my own work.

But you could pick other metrics too – they're all equivalent. The other options you're most likely to see are "precision" and "recall." Precision is the same thing as the true positive rate – the fraction of all flagged results that actually should have been flagged. Recall measures your classifier's coverage – out of all the things that *should* be flagged, it is the fraction that actually *gets* flagged.

8.6.1 Confusion Matrices

A common way to display performance metrics for a binary classifier is with a "confusion matrix." It is a 2×2 matrix displaying how many points in your testing data were placed in which category versus which category they should have been placed in. For example,

Correct label	Predicted = 0	Predicted = 1
0	35	4
1	1	10

In the given confusion matrix, the true positive rate would then be $10/(10+1) = 0.91$, and the false positive rate would be $4/(4+35) = 0.10$.

8.6.2 ROC Curves

If you treat the false positive rate as an x-coordinate and the true positive rate as the y-coordinate, then you can visualize a classifier's performance as a location in a two-dimensional box such as the following:

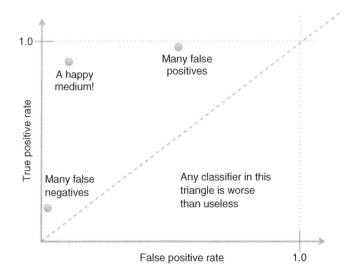

The upper-left corner (0.0, 1.0) corresponds to a perfect classifier, flagging every relevant item with no false positives. The lower-left corner means flagging nothing, and the upper-right means flagging everything. If your classifier is below the $y = x$ line, then it's worse than useless; an irrelevant item is more likely to be flagged than one that's actually relevant.

My discussion so far has been about classifiers in a binary way; they will label something as (say) fraud or nonfraud. But very few classifiers are truly binary; most of them output some sort of a score, and it's up to data scientists to pick a cutoff for what counts as a hit. This means that a single classifier is really a whole family of classifiers, corresponding to where we pick the cutoff. Each of these cutoffs corresponds to a different location in our 2d box, and together they trace out what's called an ROC curve, similar to the ones that we generated at the beginning of this chapter:

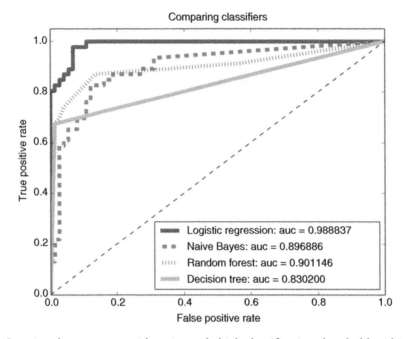

Imagine that you start with an insanely high classification threshold, so high that nothing actually gets flagged. This means that you are starting in the lower-left of the box, at (0, 0). As you loosen your criteria and start flagging a few results your location shifts. Hopefully, the first results you start flagging are all dead ringers and you get very few false positives, that is, your curve slopes sharply up from the origin. When you get to the knee of the curve, you have flagged all the low-hanging fruit and hopefully not incurred many false positives. If continue to loosen your criteria, you will start to correctly flag the stragglers, but you will also flag a lot of false positives. Eventually, everything will get flagged, and you will be at (1, 1).

If we are trying to understand the quality of the underlying score-based classifier, then it's not fair to judge it by a single threshold. You want to judge it by the entire ROC curve – a sharp "knee" jutting into the upper-left corner is the signature of a strong classifier.

8.6.3 Area under the ROC Curve

This holistic, look-at-the-whole-ROC-curve viewpoint doesn't absolve us from sometimes having to boil the performance down into a single number. Sometimes, you'll need a numerical criterion for, say, declaring that one configuration for your classifier is better than another.

The standard way to score an entire ROC curve is to calculate the area under the curve (AUC); a good classifier will mostly fill the square and have an AUC near 1.0, but a poor one will be close to 0.5. The AUC is a good way to score the underlying classifier, since it makes no reference to where we would draw a classification cutoff. When you run the numbers, there is a very clear, real-world meaning to the AUC: it is the probability that a randomly selected hit will have a higher prediction compared to a randomly selected nonhit.

In my own work, I will use the AUC to decide which of several underlying classifiers and configurations I want to use. If random forest has an AUC of 0.95, but logistic regression only has 0.85, it's clear where I'm going to focus my efforts.

In the script at the beginning of this chapter, we showed code that computes taking in our prediction scores and the correct labels and uses them to compute the FPR, TPR, classification thresholds, and AUC. The relevant lines are as follows:

```
from sklearn.metrics import roc_curve, auc
fpr, tpr, thresholds = roc_curve(Y_test, pred)
auc_score = auc(fpr, tpr
```

8.7 Selecting Classification Cutoffs

Intuitively, we want to set our thresholds so that our classifier is near the "knee" of the curve. Maybe business considerations will nudge us to one part or another of the knee, depending on how we value precision versus recall, but there are also mathematically elegant ways to do it. I'll discuss two of the most common ones in this section.

The first is to look at where the ROC curve intersects the line $y = 1-x$. This means that the fraction of all hits that end up getting flagged is equal to the fraction of all nonhits that *don't* get flagged: we have the same accuracy on the hits and the nonhits. A different cutoff would make me do better on hits but worse on nonhits, or vice versa. All other things being equal, this is the cutoff that I use, partly because I can truthfully answer the question of "how accurate is your classifier?" with a single number.

The second approach is to look at where the ROC curve has a 90% slope, that is, where it runs parallel to the line $y = x$. This is sort of an "inflection point": below this threshold, relaxing your classifier boosts the flagging probability of a hit more so than a nonhit. Above this threshold, relaxing your classifier will boost the flagging probability for nonhits more so than for hits. It's effectively like saying that an epsilon increase in TPR is worth the cost of an epsilon increase in FPR, but no more than that.

This second approach is also useful because it generalizes. You could instead decide that a tiny increase in TPR is worth three times the increase in FPR, because it's that much more important to you to find extra hits. Personally, I've never had occasion to go that far into the weeds, but you should be aware it's possible in case the need arises.

8.7.1 Other Performance Metrics

The AUC is the right metric to use when you're trying to gauge the holistic performance of a classifier that gives out continuous scores. But if you're using the classifier as a basis for making decisions, the underlying classifier score is only worth as much as the best single classifier you can make out of it. So, what you might do here is to pick a "reasonable" threshold (by the definition of your choice) for your classifier and then evaluate the performance of that truly binary classifier.

If you set your classification threshold to the point where the ROC curve intersects the line $y = 1-x$, then your classifier has the same accuracy in classifying hits that it does with nonhits. In that case, you can use this single number as the accuracy of the classifier. This has the advantage of simplicity and of being easier to explain to clients who aren't used to the idea of using two numbers to evaluate a classifier.

The classical way to judge a binary classifier is called the F_1 score. It is the harmonic mean of the classifier's precision with its recall, defined by

$$F_1 = \frac{1}{\left(\dfrac{1}{\text{Precision}} + \dfrac{1}{\text{Recall}}\right)/2}$$
$$= \frac{2 * \text{Precision} * \text{Recall}}{\text{Precision} + \text{Recall}}$$

The F_1 score will be 1.0 for a perfect classifier and 0.0 in the worst case. It's worth noting though that there is nothing magical about using the harmonic mean of precision and recall, and sometimes, you will see the geometric mean used to compute the *G*-score:

$$G = \sqrt{\text{Precision} * \text{Recall}}$$

Technically, you could even use the arithmetic mean (precision + recall)/2, but that would have the unpleasant effect that flagging everything (or not flagging anything at all) would give a score better than 0.

8.7.2 Lift–Reach Curves

Some people prefer what's called a lift–reach curve, rather than an ROC curve. It captures equivalent information, namely how the performance of a classifier varies as you adjust the classification threshold, but displays it in a different way. The lift–reach curve is based on the following notions:

- Reach is the fraction of all points that get flagged.

- Lift is the fraction of all points you flag that are hits, divided by the fraction of hits in the overall population. A lift of 1 means that you are just flagging randomly, and anything above that is positive performance.

The reach is plotted along the x-axis, and the lift along the y-axis. Typically, the lift will start off high, and then it decays to 1.0 as reach approaches 1.

8.8 Further Reading

1 Bishop, C, *Pattern Recognition and Machine Learning*, 2007, Springer, New York, NY.
2 Janert, P, *Data Analysis with Open Source Tools*, 2010, O'Reilly Media, Newton, MA.

8.9 Glossary

Area under the ROC curve The area under an ROC curve measures how well a classifier works that is independent of where the classification threshold is set.

Confusion matrix A 2×2 table giving the true positives, false positives, true negatives, and false negatives for a classifier.

Decision tree A machine learning model that works as a flowchart considering one input feature at a time.

Ensemble classifier A machine learning classifier that works by training multiple classifiers on random subsets of the rows/columns of the training data. When classifying a point, it averages the results of the different classifiers.

False positive rate The fraction of all nonhits that are errantly classified as hits.

F_1 score A measure of classifier performance. It is the harmonic mean of the precision and the recall.

Logistic regression A machine learning classifier that can be thought of as a nonbinary version of support vector machines.

Neural net A type of machine learning model inspired by human neurons.

Precision Of all flagged results flagged by a classifier, this is the fraction that are actually hits.

Random forest An ensemble of decision trees in a single classifier.

Recall The fraction of all hits that get flagged by a classifier.

ROC curve A graph that measures FPR on the x-axis and TPR on the y-axis, across all possible classification cutoffs for a classifier that outputs continuous scores.

Sigmoid function A common activation function for neural nets, whose output is always between 0 and 1.

Support vector machine A machine learning classifier that works by drawing a hyperplane in d-dimensional space and classifying points by which side they fall on.

SVM Short for "support vector machine."

True positive rate The fraction of all hits that are correctly classified as hits.

9

Technical Communication and Documentation

I debated whether to include this chapter in the book. In the first place, it ventures into "touchy feely" areas that I generally try to avoid. This is mostly a technical, brass-tacks type of book. The second problem is that I don't feel that I myself am that great at technical communication. I'm certainly good enough to get my own work done (and given that I'm in consulting, that bar is higher for me than for most data scientists), but beyond that I don't claim to have any special expertise.

However, my own lack of natural talent is part of why I felt this chapter is necessary. I've seen that first-rate technical work can be tragically undervalued if people fail to communicate it in an effective way. I've also seen that just a few basic, easy-to-learn principles can make a world of difference between incomprehensibility and a stunning presentation. Internalizing a few guiding principles has made for career advancement for myself, follow-up engagements for my company, and early identification of mismatches between the technical work and business objectives.

Data scientists are in a uniquely communication-intensive niche. Software engineers mostly talk with other software engineers, business analysts with business analysts, and so on. It is the job of a data scientist to bridge the gaps between the worlds of business, analytics, and software. So, it's a crying shame that, frankly, most of us aren't that good at it. Ultimately, everything in this chapter is about one goal: conveying ideas to your audience in a way that (1) they will actually understand and (2) they have an easy time understanding.

In this chapter I will start with a few general principles that I think underlie most good technical communication. I will then move on to specific tips for slide decks, written reports, and spoken presentations. I will also include a section on source code, which is sometimes the communication medium of last resort.

The Data Science Handbook, First Edition. Field Cady.
© 2017 John Wiley & Sons, Inc. Published 2017 by John Wiley & Sons, Inc.

9.1 Several Guiding Principles

9.1.1 Know Your Audience

This is one of the most basic principles of technical communication, but it's also one of the hardest to master. As a data scientist, you will talk to the following people:

- Domain experts who understand what you're studying better than you do but don't know much about software and analytics. You will often have to hold their hand a little bit in explaining what you did. The various real-world scenarios that got left out of your analysis are likely to be a particular focus of conversation. In my experience, these people are *extremely* useful in uncovering shortcomings in your work or making sense of issues in the data.
- Analytical people, who are very interested in the nitty-gritty of what you've done. You can expect to spend a lot of time here discussing (and possibly justifying) your statistical methodology and modeling choices.
- Software engineers, who will often want to treat your code as a black box that magically spits out answers, but they will care a lot about its performance in real-world situations. Software engineers run the gamut from barely knowing how to calculate an average to being extremely mathematically savvy.
- Business people, a diverse group, ranging from former engineers who will grill you about the details to nontechnical managers who want everything translated into business speak. The one thing they will have in common is a keen curiosity about what your work means for the different parts of the company.

Another important part of knowing your audience is knowing how much detail to include and how much of the story of your work. A high-level executive might want to know just a few key take-home points. Your peers might want more details about your methodology, especially if your findings are especially important or surprising or if you change direction mid-project because of preliminary findings. Sometimes, it might be important to go into several things you tried that didn't work, because you didn't get strong results and you need to show that it was the data's fault and not yours. In other cases, that is a waste of people's time.

9.1.2 Show Why It Matters

Always make sure to frame an analytics in the context of something people already care about, usually a business problem, in order to make it compelling. Depending on your audience, you may also need to clearly explain how the analytics relates back to the problem and can impact the bottom line. You typically don't need to belabor the point, but you should give people a reason to care about what you're saying.

Many analytics problems are clearly related to a company's business and probably don't need any motivation. Getting people to click on ads, for example, is obviously core to many business models. In other cases though, the connection is more tenuous. Do we really need customer segmentation, if we aren't currently planning to target ads? Especially in large corporations, disagreements between high-level people come up over the value of analytics projects, and chances are your boss will have to lean on you to explain why the project is worthwhile. As I write this, I'm on a project where I'm working to convince a factory manager that it's worthwhile to use analytics to study how we can reduce the bottleneck at their test bench.

This is not just about communicating with other people; it's also an excellent exercise for you. If you can't explain in simple terms why an analytics project is worthwhile, then maybe you should be working on a different problem.

9.1.3 Make It Concrete

The human brain doesn't do very well with abstract concepts. I don't just mean nontechnical people; even if somebody has the background to follow a purely abstract discussion, their understanding will be immeasurably helped if you give their brain a few concrete mental hooks.

Often, the business case at hand provides all the concrete examples you need. Other times though, the business case is too convoluted to illustrate things clearly, and you will want a simple toy problem.

9.1.4 A Picture Is Worth a Thousand Words

One of the best pieces of advice I ever got for writing papers or giving technical talks was this: the heart of your presentation is one or a few key figures. The rest of the paper is just an extended caption describing how you generated those figures and how to interpret them.

It seems like every year I decide that visuals are more important than I had previously realized, and every year I'm right. Whether it's diagrams to illustrate a concept, plots to display data, or just a stick figure scrawled on a whiteboard, this is the best way to convey ideas to another person and make them compelling.

In my opinion, the lack of pictures in some papers and presentations is often a sign of laziness (certainly, it sometimes is with mine). It takes some planning to decide what figures would work best. Then there's a lot of legwork in generating those figures, whether it's manipulating a diagram in PowerPoint or making sure that the axes are set correctly on a plot. It's a lot easier to just sit at the keyboard and churn out slides and pages of text (at least, for me it is), but that's the wrong way to go about it.

9.1.5 Don't Be Arrogant about Your Tech Knowledge

This should go without saying, but I feel compelled to bring it up because I have seen it way too often: data scientists being jerks toward people who don't know as much math as they do. Obviously, this is horrible for repeat business, and it puts up a massive barrier to clear communication. But for my two cents, I'd like to say that it's also wrong. I've seen the same data scientists who were being so arrogant go on to screw up projects, because they knew the mathematical equations but were unable to think critically about the concepts behind them.

I had a lot of fun on a consulting project once, where the manager on the client's side had no mathematical background to speak of. But whenever she asked me about the tech work, she would immediately ask all the right questions. It was almost as if I had sat down with a list of all the statistics concepts that were important to the problem, translated each one into normal English, and then put a question mark at the end. She had no training in statistics; she was just smart and level-headed enough that she zeroed in on all the right points, even better than most data scientists I've worked with. It was a fun reminder to me that math is not synonymous with clear thinking: it's just a way of reducing that clear thinking to calculations so that you can get a number out.

9.1.6 Make It Look Decent

I used to think that aesthetics was peripheral to clear communication. I felt like people should judge my work based on its technical merits, rather than how much I agonized over which shade of peach to use. So, it came as quite a shock when I first read about graphic design. I discovered that it isn't an attempt to shoehorn artistic sentiments into technical work; it is a pragmatic way to make sure that communication is clear and compelling. You should use good design principles on a slide for the same reason you should use logarithmic axes when graphing some data: it helps to get the point across.

9.2 Slide Decks

Slide decks are the most common medium for communicating data science results. They're also the medium where you can have the most impact, because people are more likely to listen to your talk or walk through a slide deck than they are to read a written report.

The biggest pitfall that I see with slide decks is people treating them as written reports. In some cases, they will go so far as to just copy and paste text from a write-up they did and call it a slide deck. This is the wrong way to do it! Slide decks are a fundamentally different communication medium and governed by a different set of rules. On the one hand, you have complete freedom to control what the slide looks like. On the other, people will expect to be able to step through a slide deck much more quickly than a report, so it is critical to keep the deck crisp, compelling, and to the point. Frequently, this means replacing textual content with graphical content wherever possible.

The second most frequent problem I see is people going overboard with what you can do in a slide deck. Crazy shapes, goofy moving images, and pictures of cats should be used very sparingly.

Ultimately, you will have to think about how best to structure your ideas around a graphical representation. This is a totally different skill set from writing or speaking in clear prose. In my experience, good graphical presentations are harder to put together compared to good prose but much easier for your audience to follow.

There is no magic bullet to making good presentations: it will require hard work on your end and making a habit of asking whether something could be explained a little more clearly. To get you started though, a good rule of thumb is that 75% of your slides should consist of an image (a data graphic, a flowchart, etc.) and at most two phrases/sentences that give the take-home message for that image.

9.2.1 C.R.A.P. Design

A lot of the principles of good design are captured in the acronym C.R.A.P, which stands for contrast, repetition, alignment, and proximity:

- Contrast: Things that are different should look different. This makes it seamless for people to notice and internalize the differences. For example:
    - Use different fonts for code, text, and figure captions.
    - Use different colors for different customer segments that you've identified.
    - Within reason, use different font sizes on slides to emphasize different points.
- Repetition: Key points or design motifs should be repeated throughout your work. People are likely to miss them the first time, so you want to repeat them in a way that is obvious enough to have an impact but subtle enough that they're not annoying if people already got the point. Repetition is partly the dual of contrast: if things are similar or related, or if they are different, those patterns should be carried through consistently.
- Alignment: Of all the principles, this comes the closest to being a purely aesthetic thing. Make sure that the different parts of your visual field line up with each other in a natural way. Or, if they shouldn't line up (maybe because you're trying to contrast them), then make sure they are obviously not aligned. The last thing you want is for somebody to be distracted during a talk by wondering whether two blocks of text are actually slightly out of line.

- Proximity: Use distance between things to indicate their relationships. More generally, use the layout of the visual field to your advantage.

To show you just how bad it can get when the basic principles of design are brazenly flaunted, I humbly present one of my own slides that I dug up from back in the day:

Overview

- Purpose
  - Numerical and mathematical computing for Python
  - Make it FAST
- NumPy
  - Core extension to Python
  - Support for n-dimensional arrays
  - Mathematical operations on arrays
- SciPy
  - Extensive libraries for technical computation
  - Operates on NumPy arrays

Among the problems in this slide, which I somehow managed to stand up in front of people and present, are the following:

- Almost everything is in the same font.
- The purpose of the talk looks like it's just another topic, on par with SciPy and NumPy.
- There is an unsightly amount of white space.
- Everything is aligned with everything else.

If I could take back that presentation, I would replace the slide with something like the one on the next page:

The changes I wish I could make include the following:

- [Contrast] The slide title, goal, topic text all look different.
- [Repetition] Those text differences are marked consistently.
- [Alignment] The NumPy and SciPy sections line up with each other, as do their supporting pictures.
- [Alignment] NumPy and SciPy, the two subjects of the talk, both have their own half of the space. The stuff that applies to the whole slide is in the middle.
- [Proximity] The word "NumPy," the description of NumPy, and the suggestive pictures are all next to each other. Similarly for SciPy. The goal line is next to the slide title.

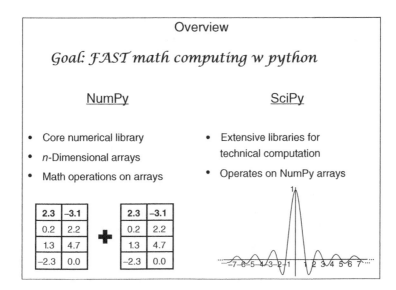

- I added some suggestive pictures. Realistically, many people aren't going to read my descriptions of NumPy and SciPy. But even so, they'll still get the idea that NumPy does basic operations on arrays, while SciPy does fancier mathematical operations.

The content of the slides is the same. Graphic design is there to grease the wheels of communication. It's not there to supplement my content: it's there to make you notice it.

9.2.2 A Few Tips and Rules of Thumb

Making good slides is a skill that only develops with practice. But to help you avoid a few common pitfalls, here are several helpful hints or rules of thumb that have served me well:

- Avoid long sentences or phrases. You can often replace a complete sentence with a shorter phrase.
- Have no more than four bullet points in any given section.
- If you feel like there's too much you want to fit into a slide, it's fine to break it out into two slides.
- If you can replace a bulleted list with something else, such as callouts with arrows pointing at different parts of a figure, you should probably do it.
- If the same image or section occurs in the same place on two consecutive slides, make absolutely sure that it is in *exactly* the same place. Nothing is more visually grating than having something shift over by a few pixels when you go from one slide to the next.

- Avoid the (in)famous Comic Sans MS font altogether unless you're sure about your audience. Personally, I think comic sans is useful when employed judiciously, but a lot of people feel strongly that it looks unprofessional and should just never be used. Know your audience.
- PNG files are your friend, because they can have transparent backgrounds. This lets you put an image in your slide without having its white background mar other images or the slide's backdrop. Personally, I keep a slide deck consisting of only images with transparent backdrops (logos for different programming languages, etc.) to save myself the trouble of looking for such images on the Internet each time I need them.
- Make sure the axes are labeled on all of your figures. I made this mistake once... not good.
- Have a slide background. This is easy to do and makes slides look so much better. Keep it simple, so that it doesn't distract people from your content.
- Make sure that your figures include some color if possible.
- If you want two different colors in a color scheme, don't use red and green. Many people are partly colorblind and can't see the difference.
- It's fine to be more casual in a slide deck than you would in a written report. But don't overdo it.
- When planning your presentation schedule, the rule of thumb that people use is that one slide will take about 2 minutes to cover.
- In some cases, you might want to include slide numbers, dates, or a company logo or something on every slide. It's good to give people concrete reference points.

9.3 Written Reports

First off, please don't use LaTeX unless you are planning to publish your work in a scientific journal or you are deliberately trying to look academic so as to make an impression on somebody. This is one of my pet peeves, which I see distressingly often since so many data scientists come from an academic background where LaTeX is standard. If you haven't heard of it, LaTeX is a markup language that can be compiled into beautifully formatted documents, and it is very popular for publishing scientific papers. The downside though is that you need to know the idiosyncrasies of LaTeX's syntax in order to edit the document, which bars most people from editing it collaboratively. It's true that LaTeX gives you very fine-grained control over what a document looks like, but that power is rarely necessary and often abused. I generally recommend that you use Microsoft Word, Google Docs, or some other WYSIWYG editor. I'm writing this book in Word.

Now that that's out of the way, let's talk about content and presentation. The structure of a written report will vary depending on your intended audience, who you are in relation to them (team member, outside consultant, member of another team, etc.), and the problem you're addressing. However, most technical reports will have some subset of the following sections:

- An executive summary. This is up to one page that summarizes what problem you were addressing and why, what you did, and what can be done with it. The emphasis should be on the takeaway points from a business perspective and how your work fits into the larger context of a company.
- Background and motivation. Clearly, frame how this work fits into a larger context for your likely audience. Depending on who you're writing for, it might be a description of how this fits into the company's business, the role it plays in software, or existing knowledge that it builds on.
- Datasets used. Describe in brief which datasets are being used, where you got them from, and what they're describing. Plus maybe a little bit about which features you extract from them and any limitations of the data that should be pointed out. This section should be short and sweet; if there are a lot of gory details, put them in an appendix.
- Analytical overview. Describe at a high level the analysis you performed or the algorithm you are studying. Focus on the mathematical model in the abstract, rather than how it is implemented in software (unless some key aspects of it were driven by software requirements, such as wanting it to be massively parallel). This section should probably have a diagram or two that illustrate what you're talking about.
- Results. Describe any results you got from your analysis, and present them in graphical form. This is often the most important part of your report, so keep it crisp and compelling, and make sure to tie it back to the context of how these results are relevant. If you have a lot of results to report that contain similar information (such as results for each feature), then include only the most interesting ones in this section. Put the rest in an appendix.
- Software overview. This section often doesn't need to be there and should be short if you do include it. It's mostly relevant if your code is being plugged into somebody else's code or some production system or if your code might be regularly rerun in the future as datasets are updated. Describe how to run the code (this should be at most a handful of lines – if it's not, then you should refactor it and maybe combine it into a master script) if it's a stand-alone analysis or how it plugs into other software if that's how it works. Include a high-level architecture of the code as a diagram, and describe which languages it is written in and what tools it uses.
- Future work. Discuss natural next steps. This section often reads as boilerplate in practice, and sometimes, it's ok if it's extremely brief or even omitted entirely. However, it can also be an opportunity to point the way to

significant new projects and to suggest others that should not be pursued. Data science is often used to "test the waters" and see whether something is worth pursuing as a larger project or clarify the scope that such a project should have.

- Conclusions.
- Appendices with technical details. For me, personally, up to half my report is liable to be appendices.

9.4   Speaking: What Has Worked for Me

I have never benefitted from the classic technique of imagining your audience naked.

More seriously though, different people have very different styles of presentation. Some people script out their presentations and practice them down to the word, making sure that every inflection is carefully chosen but still feels natural. Personally, I've never had much success with that approach. What works best for me is to:

- Make sure I've got a great slide deck.
- Practice a lot, discussing each slide in the deck and the key concepts it's discussing. Not scripting them out, but giving clear explanations out loud.
- When it comes time for the actual presentation, throw all my practice out the window and wing it. I keep half an eye on time, but otherwise just explain my slides in whatever way feels most natural.

This is what works for me. I tend to speak very well spontaneously, and many of my best talks are given at the spur of a moment, on topics that I know well but wasn't planning to discuss. But when I have an actual plan for what I'm going to say, I immediately become mumbly and boring.

The gold standard of presentations is to make it feel like a conversation to your audience (with a well-composed, clearly-thought-out person who is good at explaining things) rather than any kind of formal speech. There is an awful lot that goes into making it look natural: tone of voice, cadence, facial expressions, and moving your body, all of which are independent of the actual words you are saying. This is too much to keep track of while you are focusing on your actual talk, so the only solution is to practice so that these things become second nature.

I recommend training yourself to have a "speaking personality," a game face that you can put on at will. You should regularly practice giving off-the-cuff explanations of things that are in your line of work. This might be explaining how a cache works, describing the goals and status of a project you're working on, or giving an elevator spiel about your grad school research. It can be done in front of a mirror, in front of friends, or even just in your own head as an internal monologue. During your explanations, focus on your content but also keep an eye toward the following guidelines:

- Be clear and enunciate.
- Speak a little more slowly than you would in normal conversation. People will be paying attention to your slides as well as your words, and you won't be able to repeat something if a single person misses it. So, give everybody a little more time to absorb what you're saying. Plus, this helps if (like me) you tend to speak faster when you're nervous.
- Throwing in a short personal anecdote, joke, or opinion can make your presentation more relatable and interesting. Don't go overboard, but this can add a nice humanizing touch to technical material.
- A short pause is always better than "um."
- Let yourself be animated, so that it's clear you're excited about what you're discussing. Add personality!
- Adopt a natural, at-ease cadence. Even if somebody isn't paying attention to your talk, they will immediately notice anxiety or nervousness in your voice.
- Try to have good posture: stand up straight, hold your head high, and gently pull your shoulders back and down. This little bit of polish doesn't just make you look more confident; research has shown that adopting confident body language makes you more confident in reality.
- Keep an eye toward your hand movements. Some people move their hands nervously in a way that is very distracting. But moderate use of hand gestures can add emphasis and personality.

With enough practice, these finer points will become second nature. You will be able to easily shift in and out of "talk mode" similarly to an actor going in and out of character. At that point, it won't just feel natural; it will now be second nature for you.

Many people have a deep-seated anxiety about public speaking. If you are in that situation, then chances are this section won't give you everything you need. I encourage you to look into Toast Masters or a similar group that helps people practice public speaking and learn to feel at ease with it.

9.5 Code Documentation

Whenever you provide a significant piece of code as a deliverable, it's important to provide some kind of documentation of what it does and how to use it. Depending on the context that can take a variety of forms, including the following:

- A long comment at the top of a file.
- A separate runbook or user manual. This is more common with extremely large pieces of software or if you're giving it to a client or another team.

- Pages on a company Wiki.
- Unit tests that can be run against the code.

No matter what form the documentation is in though, the most important thing about the documentation is to explain how somebody can run the code and reproduce its functionality. This lets them use the code themselves and verify that it works the way that it's supposed to.

Explaining how it operates under the hood is secondary, and going into too much detail can be counterproductive. If you are delivering your source code to somebody, it is generally reasonable to expect them to be able to read and understand your code, and restating it all in English is superfluous. Telling them how to run the software tells them where to start looking in the source code, and they can follow the thread from there. It is good to give a brief architectural overview, pointing out which modules do what, but I wouldn't go beyond that.

The one thing that is very nice to include though is a troubleshooting section. Most pieces of software have some weird ways they can break down, or parts that are known to be especially fragile, that are highly specific to your software. If this is the case, save your users potentially hours of time debugging by telling them what they probably did wrong.

9.6 Further Reading

Kolko, J, *Exposing the Magic of Design*, 2011, Oxford University Press,
New York, NY
Matplotlib 1.5.1 Documentation, viewed 7 August 2016, http://matplotlib.org/

9.7 Glossary

Comic sans A font that is notorious for being very casual, some would argue overly so.

C.R.A.P design The idea that you should use contrast, repetition, alignment, and proximity as underlying principles in graphic design.

LaTeX A markup language that is a very popular tool for publishing scientific papers. It can be compiled into beautifully formatter papers, but it is finicky to use and has a steep learning curve.

Runbook A document explaining what a piece of code does, how to use it, and how to troubleshoot it.

Part II

Stuff You Still Need to Know

The second section of this book will cover a variety of topics that are liable not to show up in a given data science project. This doesn't mean that knowing them is optional for a professional data scientist, but it does mean that you might not be put on the spot about them until somewhat later in your career. My goal is to fill those holes ahead of time.

The topics here cover a wide range. There are very general-purpose analytics tools that almost made it into the first part, such as clustering. Most software engineering concepts, beyond basic scripting, fit into this chapter was well. Finally, there are very specialized areas such as natural language processing, which some data scientists never use at all.

The Data Science Handbook, First Edition. Field Cady.
© 2017 John Wiley & Sons, Inc. Published 2017 by John Wiley & Sons, Inc.

10

Unsupervised Learning: Clustering and Dimensionality Reduction

This chapter is about techniques for studying the latent structure of your data, in situations where we don't know a priori what it should look like. They are often called "unsupervised" learning because, unlike classification and regression, the "right answers" are not known going in. There are two primary ways of studying a dataset's structure: clustering and dimensionality reduction.

Clustering is an attempt to group the data points into distinct "clusters." Typically, this is done in the hopes that the different clusters correspond to different underlying phenomena. For example, if you plotted people's height on the x-axis and their weight on the y-axis, you would see two more-or-less clear blobs, corresponding to men and women. An alien who knew nothing else about human biology might hypothesize that we come in two distinct types.

In dimensionality reduction, the goal isn't to look for distinct categories in the data. Instead, the idea is that the different fields are largely redundant, and we want to extract the real, underlying variability in the data. The idea is that your data is d-dimensional, but all of the points *actually* only lie on a k-dimensional subset of the space (with $k < d$), plus some d-dimensional noise. For example, in 3d data, your points could line mostly just along a single line or perhaps in a curved circle. Real situations of course are usually not so clean cut. It's more useful to think of k dimensions as capturing "most" of the variability in the data, and you can make k larger or smaller depending on how much of the information you want to reproduce.

A key practical difference between clustering and dimensionality reduction is that clustering is generally done in order to reveal the structure of the data, but dimensionality reduction is often motivated mostly by computational concerns. For example, if you're processing sound, image, or video files, d is likely to be tens of thousands. Processing your data then becomes a massive computational task, and there are fundamental problems that come with having more dimensions compared to data points (the "curse of dimensionality," which I'll

The Data Science Handbook, First Edition. Field Cady.
© 2017 John Wiley & Sons, Inc. Published 2017 by John Wiley & Sons, Inc.

discuss shortly). In these cases, dimensionality reduction is a prerequisite for almost any analytics you might want to do, regardless of whether you're actually interested in the data's latent structure.

10.1 The Curse of Dimensionality

Geometry in high-dimensional spaces is weird. This is important, because a machine learning algorithm with d features operates on feature vectors that live in d-dimensional spaces. d can be quite large if your features are, say, all of the pixel values in an image! In these cases, the performance of these algorithms often starts to break down, and this decay is best understood as a pathology of high-dimensional geometry, the so-called curse of dimensionality.

The practical punch line to all of this is that if you want your algorithms to work well, you will usually need some way to cram your data down into a lower-dimensional space. There's no need to dwell too much on the curse of dimensionality – thinking about it can hurt the heads up us three-dimensional beings.

But if you're interested, I would like to give you at least a tiny taste of what goes on in high dimensions. Basically, the problem is that in high dimensions, different points get very far away from each other. To illustrate, the following code lets us set d as a parameter. It then generates a thousand random points in the unit cube, calculates the distance from every point to every other point, and shows a histogram of those distances for $d = 2$ and $d = 500$.

```
import numpy, scipy
d = 500
data = numpy.random.uniform(
    size=d*1000).reshape((1000,d))
distances = scipy.spatial.distance.cdist(data, data)
pd.Series(distances.reshape(1000000)).hist(bins=50)
plt.title("Dist. between points in R%i" % d)
plt.show()
```

You can see that for $d = 500$, two points in the cube are almost always about the same distance from each other. If you did a similar simulation with spheres, you would see that almost all the mass of a high-dimensional sphere is in its crust.

Sounds weird? Well yes, it is. That's why we reduce dimensions.

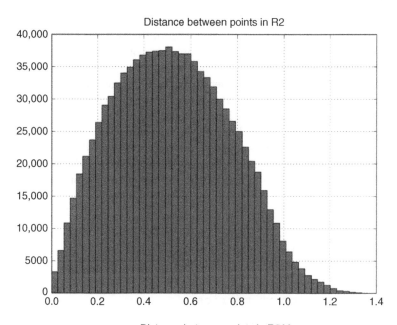

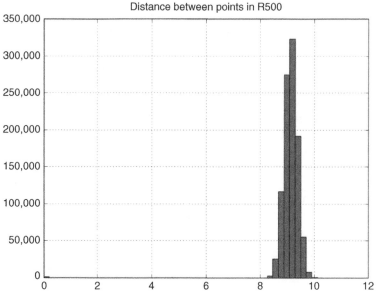

10.2 Example: Eigenfaces for Dimensionality Reduction

The following script will dive right into a lot of the material that we will cover in this chapter. It loads in a sample data of images of 64×64 pixel faces, with 10 pictures each of 40 different people. That's $d = 64*64 = 4096$ dimensions. It then clusters the images, prints out a measure of how distinct the clusters are from each other, and then prints out a measure of how closely the identified clusters line up with the identities of the humans pictured.

Then the script uses a technique called "principal component analysis" (which we will go over in this chapter) to reduce the 4096-dimensional images down to a saner 25 dimensions and redoes the analysis. It finds that the identified clusters match up slightly better with the humans being pictured.

```
import sklearn
import sklearn.datasets as datasets
from sklearn.decomposition import PCA
from sklearn.cluster import KMeans
from sklearn.metrics import silhouette_score,
adjusted_rand_score
from sklearn import metrics

# Get data and format it
faces_data = datasets.fetch_olivetti_faces()
person_ids, image_array = faces_data['target'], faces_
data.images
# unroll each 64x64 image -> (64*64) element vector
X = image_array.reshape((len(person_ids), 64*64))

# Cluster raw data and compare
print "** Results from raw data"
model = KMeans(n_clusters=40)
model.fit(X)
print "cluster goodness: ", silhouette_score(X, model.
labels_)
print "match to faces: ", metrics.adjusted_rand_score(
    model.labels_, person_ids)   # 0.15338

# Use PCA to
print "** Now using PCA"
pca = PCA(25)   # pass in number of components to fit
pca.fit(X)
X_reduced = pca.transform(X)
model_reduced = KMeans(n_clusters=40)
model_reduced.fit(X_reduced)
```

```
labels_reduced = model_reduced.labels_
print "cluster goodness: ", \
    silhouette_score(X_reduced, model_reduced.labels_)
print "match to faces: ", metrics.adjusted_rand_score(
    model_reduced.labels_, person_ids)
```

When I run this script, my output is

```
** Results from raw data
cluster goodness:   0.148591
match to faces:   0.454254676789
** Now using PCA
cluster goodness:   0.230444
match to faces:   0.467292493785
```

In the interest of visualizing the data itself, you can continue the analysis with the following script in order to get better insight into the PCA process. It shows a picture of one of the raw images and then displays the first two so-called eigenfaces that were identified. PCA tries to model every picture in the dataset as a mixture of the most important eigenfaces, so visualizing them can give us an idea of how the faces vary across the dataset. Finally, it shows a graph called a "skree plot," which plots out the importance of the different eigenfaces.

```
# Display a random face, to get a feel for the data
sample_face = image_array[0,:,:]
plt.imshow(sample_face)
plt.title("Sample face")
plt.show()
# Show eigenface 0
eigenface0 = pca.components_[0,:].reshape((64,64))
plt.imshow(eigenface0)
plt.title("Eigenface 0")
plt.show()
eigenface1 = pca.components_[1,:].reshape((64,64))
plt.imshow(eigenface1)
plt.title("Eigenface 1")
plt.show()
# Skree plot
pd.Series(
    pca.explained_variance_ratio_).plot()
plt.title("Skree Plot of Eigenface Importance")
plt.show()
```

The script generates the following figures:

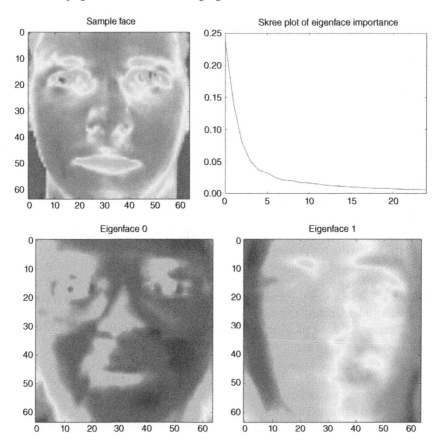

10.3   Principal Component Analysis and Factor Analysis

The granddaddy of dimensionality reduction algorithms is, without question, principal component analysis or PCA.

Geometrically, PCA assumes that your data in d-dimensional space is "football shaped" – an ellipsoidal blob that is stretched out along some axes, narrow in others, and generally free of any massive outliers. Take the following image, for example:

Intuitively, the data is "really" one-dimensional, lying on the line $x = y$, but there is some random noise that slightly perturbs each point. Rather than

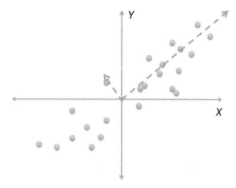

giving the two features x and y for each point, you can get a good approximation with just the single feature $x + y$. There are two ways to look at this:

- Intuitively, it seems that $x + y$ might be the real-world feature underlying the data, while x and y are just what we measured. By using $x + y$, we are extracting a feature that is more meaningful than any of our actual raw features.
- Technically, it is more computationally efficient to process one number rather than two. If you wanted, in this case, you could estimate x and y pretty accurately if you only knew $x + y$. Numerically, using one number rather than two lets us shy away from the curse of dimensionality.

PCA is a way of (1) identifying the "correct" features such as $x + y$ that capture most of the structure of the dataset and (2) extracting these features from the raw data points.

To be a little more technical, PCA takes in a collection of d-dimensional vectors and finds a collection of d "principal component" vectors of length 1, called p_1, p_2, ... and p_d. A point x in the data can be expressed as

$$x = a_1 p_1 + a_2 p_2 + ... + a_d p_d$$

However, the p_i are chosen so that generally a_1 is much larger than the other a_i, a_2 is larger than a_3 and above, etc. So realistically, the first few p_i capture most of the variation in the dataset, and x is a linear combination of the first few p_i and some small correction terms. The ideal case for PCA is something where large swaths of features tend to be highly correlated, such as a photograph where adjacent pixels are likely to have similar brightness. So, our example script was an excellent candidate.

The code in our script that performed the PCA analysis and reduced the dataset's dimension is

```
pca = PCA(25)
pca.fit(X)
X_reduced = pca.transform(X)
```

Note that in this case, we have passed the number of components we want to extract as a parameter in PCA. Under the hood, it's much more computationally efficient to only extract the first few components, so that's good to do if you don't need entire PCA decomposition.

10.4   Skree Plots and Understanding Dimensionality

In our motivation for PCA, we suggested that the dataset was "really" one-dimensional and that one of the goals of PCA is to extract that "real" dimensionality of a dataset by seeing how many components it took to capture most of the dataset's structure. In reality, it's rarely that clear. What you get instead is that the first few components are the most important, and then there is a gentle taper into uselessness as you go further out.

It is common to plot out the importance of the different principal components (technically, the importance is measured as the fraction of a dataset's variance that the component accounts for) in what's called a "skree plot," which we generated earlier:

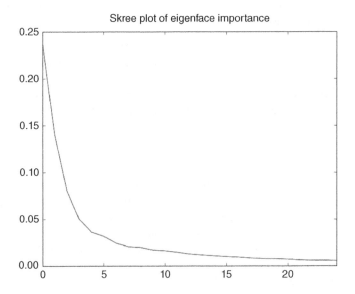

Eyeballing this plot, I would venture that the face dataset is more or less 15 dimensions – still a lot, but much less than the 4096 dimensions in the raw data.

10.5 Factor Analysis

I should note that PCA is related to the statistical technique of factor analysis. Mathematically, they're the same thing: you find a change of coordinates where most of the variance in your data exists in the first new coordinate, the second most in the second coordinate, and so on. The divergent terminology is mostly an accident of the fact that they were developed independently and applied in very different ways to different fields.

In PCA, the idea is generally dimensionality reduction; you find how many of these new coordinates are needed to capture most of your data's variance, and then you reduce your data points to just those few coordinates. A prototypical application of PCA is analyzing pictures of faces: there is a staggering number of dimensions in the data, which are almost entirely redundant, and examining the principal components themselves gives insights into the behavior of the system.

Factor analysis, on the other hand, is more about identifying causal "factors" that give rise to the observed data. A major historical example is in the study of intelligence. Tests of intelligence in many different areas (math, language, etc.) were seen to be correlated, suggesting that the same underlying factors could affect intelligence in many different areas. Researchers found that a single so-called g-factor accounts for about half of the variance between different intelligence tests. This book generally looks at things from a PCA perspective, but you should be aware that both viewpoints are useful.

10.6 Limitations of PCA

There are three big gotchas when using PCA:

- Your dimensions all need to be scaled to have comparable standard deviations. If you arbitrarily multiplied one of your features by a thousand (maybe by measuring a distance in millimeters rather than in meters, which in principle does not change the actual content of your data), then PCA will consider that feature to contribute much more to the dataset's variance. If you are applying PCA to image data, then there is a good chance that this isn't a big deal, since all pixels are usually scaled the same. But if you are trying to perform PCA on demographic data, for example, you have to have somebody's income and their height measured to the same scale. Along with this limitation is the fact that PCA is very sensitive to outliers.
- PCA assumes that your data is linear. If the "real" shape of your dataset is that it's bent into an arc in high-dimensional space, it will get blurred into several principal components. PCA will still be useful for dimensionality reduction, but the components themselves are likely not to be very meaningful.

- If you are using PCA on images of faces or something similar, the key parts of the pictures need to be aligned with each other. PCA will be of no use if, for example, the eyes are covered by different pixels. If your pictures are not aligned, doing automatic alignment is outside the skill set of most data scientists.

10.7 Clustering

Clustering is a bit of a dicier issue than using PCA. There are several reasons for this, but many of them boil down to the fact that it's clear what PCA is supposed to do, but we're usually not quite sure what we want out of clustering. There is no crisp analytical definition of "good" clusters; every candidate you might suggest has a variety of very reasonable objections you could raise. The only real metric is whether the cluster reflects some underlying natural segmentation, and that's very hard to assess: if you already know the natural segments, then why are you clustering?

To give you an idea of what we're up against, here are some of the questions to keep in the back of your mind

- What if our points fall on a continuum? This fundamentally baffles the notion of a cluster. How do I want to deal with that?
- Should clusters be able to overlap?
- Do I want my clusters to be nice little compact balls? Or, would I allow something such as a doughnut?

10.7.1 Real-World Assessment of Clusters

I will talk later about some analytical methods for assessing the quality of the clusters you found. One of the most useful of those is the Rand index, which allows you to compare your clusters against some known ground truth for what the clusters ought to be. That's what I used in the aforementioned sample script.

Usually though, you don't have any ground truth available, and the question of what exactly constitutes a good cluster is completely open-ended. In this case, I recommend a battery of sanity checks. Some of my favorite include the following:

- For each cluster, calculate some summary statistics of it based on features that were NOT used as input to the clustering. If your clusters really correspond to distinct things in the real world, then they should differ in ways that they were not clustered on.
- Take some random samples from the different clusters and examine them by hand. Do the samples from different clusters seem plausibly different?

- If your data is high-dimensional, use PCA to project it down to just two dimensions and do a scatterplot. Do the clusters look distinct? This is especially useful if you were able to give a real-world interpretation to the principal components.
- Screw PCA. Pick two features from the data that you care about and do a scatterplot on just those two dimensions. Do the clusters look reasonable?
- Try a different clustering algorithm. Do you get similar clusters?
- Redo the clustering on a random subset of your data. Do you still get similar clusters?

Another big thing to keep in mind is whether it is important to be able, in the future, to assign new points to one of the clusters we have found. In some algorithms, there are crisp criteria for which cluster a point is in, so it's ease to label new points that we didn't train on. In other algorithms though, a cluster is defined by the points contained in it, and assigning a new point to a cluster requires reclustering the entire dataset (or doing some kind of a clever hack that's tantamount to that).

10.7.2 *k*-Means Clustering

The *k*-means algorithm is one of the simplest techniques to understand, implement, and use. It starts with vectors in *d*-dimensional space, and the idea is to partition them into compact, nonoverlapping clusters. I say again: the presumed clusters are compact (not loops, not super elongated, etc.) and not overlapping.

The classical algorithm to compute the clusters is quite simple. You start off with *k*-cluster centers, then iteratively assign each data point to its closest cluster center, and recompute the new cluster centers. Here is the pseudocode:

```
1. Start off with k initial cluster centers.
2. Assign each point to the cluster center that it's
closest to.
3. For each cluster, recompute its center as the
average of all its assigned points.
4. Repeat 2 and 3 until some stopping criterion is met.
```

There are clever ways to initialize the clusters if they are not present at the beginning and to establish when they have become stable enough to stop, but otherwise the algorithm is very straightforward.

In scikit-learn, the code to perform the clustering looks as the following:

```
from sklearn.cluster import KMeans
model = KMeans(n_clusters=k)
model.fit(my_data)
```

```
labels = model.labels_
cluster_centers = model.cluster_centers_
```

k-Means clustering has an interesting property (sometimes a benefit) that when *k* is larger than the "actual" number of clusters in the data, you will split a large "real" cluster into several computed ones. In this case, *k*-means clustering is less of a way to identify clusters and more of a way to partition your dataset into "natural" regions, as shown in this figure where we set *k* = 3, but there were really two clusters:

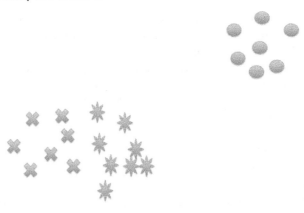

Situations such as this can often be found using the silhouette score to find clusters that are not very distinct.

The results of *k*-means clustering are extremely easy to apply to new data; you simply compare a new data point to each of the cluster centers and assign it to the one that it's closest to.

An important caveat about *k*-means is that there is no guarantee about finding optimal clusters in any sense. For this reason, it is common to restart it several times with different, random initial cluster centers. Scikit-learn does this by default.

10.7.3 Gaussian Mixture Models

A key feature of most clustering algorithms is that every point is assigned to a single cluster. But realistically, many datasets contain a large gray area, and mixture models are a way to capture that.

You can think of Gaussian mixture models (GMMs) as a version of k-means that captures the notion of a gray area and gives confidence levels whenever it assigns a point to a particular cluster.

Each cluster is modeled as a multivariate Gaussian distribution, and the model is specified by giving the following:

1) The number of clusters

2) The fraction of all data points that are in each cluster
3) Each cluster's mean and its *d-by-d* covariance matrix.

When training a GMM, the computer keeps a running confidence level of how likely each point is to be in each cluster, and it never decides them definitively: the mean and standard deviation for a cluster will be influenced by every point in the training data, but in proportion to how likely they are to be in that cluster. When it comes time to cluster a new point, you get out a confidence level for every cluster in your model.

Mixture models have many of the same blessings and curses of *k*-means. They are simple to understand and implement. The computational costs are very light and can be done in a distributed way. They provide clear, understandable output that can be easily used to cluster additional points in the future. On the other hand, they both assume more-or-less convex clusters, and they are both liable to fall into local minimums when training.

In scikit-learn, the code to fit a GMM looks such as the following:

```
from sklearn import mixture
model = mixture.GMM(n_components=5)
model.fit(my_data)
cluster_means = model.means_
# array giving the weight of each datapoint for each
of the clusters
labels = model.predict_proba(my_date)
```

I should also note that GMMs are the most popular instance of a large family of mixture models. You could equally well have used something other than Gaussians as the models for your underlying clusters or even had some clusters be Gaussians and others something else. Most mixture model libraries use Gaussian, but under the hood, they are all trained with something called the EM algorithm, and it is agnostic to the distribution being modeled.

10.7.4 Agglomerative Clustering

Hierarchical clustering is a general class of algorithms sharing a common structure. We start off with a large number of small clusters, typically with each point being its own cluster. We then successively merge clusters together until they form a single giant cluster. So, the output isn't a single clustering of the data, but rather a hierarchy of potential clusterings. How you choose which clusters to merge and how we find the "happy medium" clustering in the hierarchy determine the specifics of the algorithm.

An advantage of hierarchical clustering over k-means is that (depending on how you choose to merge your clusters) your clusters can be of any size or shape. A disadvantage though is that there's no natural way to assign a new point to an existing cluster.

The following Python code will perform an agglomerative clustering of the data until five clusters remain, merging, and assigning each point to one of the clusters. "Ward" linkage means that we pick which clusters to merge by looking for the two clusters with the lowest variance between them.

```
from sklearn.cluster import AgglomerativeClustering
clst = AgglomerativeClustering(
    n_clusters=5, linkage='ward')
cluster_labels = clst.fit_predict(my_data)
```

10.7.5 Evaluating Cluster Quality

Algorithmic methods to evaluate the outcome of clustering come in two major varieties. First, there are the supervised ones, where we have some ground-truth knowledge about what the "right" clusters are, and we see how closely the clusters we found match up to them. Then, there are the unsupervised ones, where we think of the points as vectors in d-dimensional space and look at how geometrically distinct the clusters are from each other.

10.7.6 Silhouette Score

Silhouette scores are the most common unsupervised method you'll see, and they are ideal for scoring the output of k-means clustering. It is based on the intuition that clusters should be dense and widely separated from each other, so similar to k-means, it works best with dense, compact clusters that are all of comparable size. Silhouette scores aren't applicable to things such as a dough-nut-shaped cluster with a compact one in the middle.

Specifically, every point is given a "silhouette coefficient" defined in terms of

- a = the average distance between the point and all other points in the same cluster
- b = the average distance between the point and all other points in the next-closest cluster.

The silhouette coefficient is then defined as

$$s = \frac{b-a}{\max(a,b)}$$

The coefficient is always between -1 and 1. If it is near 1, this means that b is much larger than a, that is, the point is on average much closer to points in its own cluster. A score near 0 suggests that the point is equidistant from the two clusters, that is, they overlap in space. A negative score suggests that the point is wrongly clustered.

The silhouette score for a whole cluster is the average coefficient over all points in the cluster.

The silhouette score is far from perfect. For example, imagine that one cluster is much larger than another and that they are very close, as in the following figure. The point indicated is clearly in the right cluster, but because its cluster is so large, it is far away from most of its clustermates. This will give it a poor silhouette score because it is closer, on average, to the points in the nearby cluster than its own cluster.

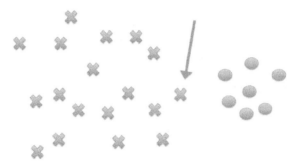

However, the silhouette score is straightforward to compute and easy to understand, and you should consider using it if your cluster scheme rests on similar assumptions.

The silhouette score is built into scikit-learn and can be called in the following way:

```
from sklearn.metrics import silhouette_score
from sklearn.metrics import silhouette_samples
coeffs_for_each_point = silhouette_samples(mydata,
labels)
avg_coeff = silhouette_score(mydata, labels)
```

10.7.7 Rand Index and Adjusted Rand Index

The Rand index is useful when we have knowledge of the correct clusters for at least some of the points. It doesn't try to match up the clusters we found to the right ones we know. Instead, it is based on the idea of whether two points that should be in the same cluster are, indeed, in the same cluster.

A pair of points (x, y) is said to be "correctly clustered" if we put x and y in the same cluster and our ground truth also has them in the same cluster. The pair is also correctly clustered if we put x and y in different clusters and our ground truth has them in different clusters. If there are n points for which we know the right cluster, there are $n(n-1)/2$ different pairs of such points. The Rand index is the fraction of all such points that are correctly clustered. It ranges from 0 to 1, with 1 meaning that every single pair of points is incorrectly clustered.

The problem with the Rand index is that even if we cluster points randomly, we will get some pairs correct by chance and have a score greater than 0. In fact, the average score will depend on the relative sizes of the correct clusters; if there are many clusters and they're all small, then most pairs of points will, by dumb luck, be correctly assigned to different clusters.

The "adjusted Rand index" solves this problem by looking at the sizes of the identified clusters and the size of the ground-truth clusters. It then looks at the range of Rand indices possible given those sizes and scales the Rand index so that it is 0 on average if the cluster assignments are random and still maxes out at 1 if the match is perfect. Code for the rand index is as follows:

```
from sklearn import metrics
labels_true = [0, 0, 0, 1, 1, 1]
labels_pred = [0, 0, 1, 1, 2, 2]
metrics.adjusted_rand_score(labels_true, labels_pred)
```

10.7.8 Mutual Information

Another supervised cluster quality metric is mutual information. Mutual information is a concept from information theory and is similar to correlation, except that it applies to categorical variables instead of numerical ones. In the context of clustering, the idea is that if you pick a random data point from your training data, you get two random variables: the ground-truth cluster that the point should be in and the identified cluster that it was assigned to. The question is how good a guess you can make about one of these variables if you only know the other one. If these probability distributions are independent, then the mutual information will 0, and if either can be perfectly inferred from the other, then you get the entropy of the distribution.

The mutual information score is available in scikit-learn as the following:

```
from sklearn import metrics
labels_true = [0, 0, 0, 1, 1, 1]
labels_pred = [0, 0, 1, 1, 2, 2]
metrics.mutual_info_score(labels_true, labels_pred)
```

10.8 Further Reading

1 Bishop, C, *Pattern Recognition and Machine Learning*, 2007, Springer, New York, NY.
2 *Scikit-learn 0.171.1 documentation*, http://scikit-learn.org/stable/index.html, viewed 7 August 2016, The Python Software Foundation.

10.9 Glossary

Adjusted Rand index A variant of the Rand index that is 0 when the that is 0 when the Rand index is no better than chance.
Agglomerative clustering A clustering method where you look for two clusters (which could be just single points) that are close to each other and then merge them into a single cluster. You do this until some stopping criterion is reached.
Eigenface When using PCA on images of faces, the eigenfaces are the principal components. When viewed as images, they are usually ghostly pseudo-faces.
Gaussian mixture model A mixture model where all clusters are modeled as Gaussian distributions.
k-means clustering Probably, the most popular clustering method. It assumes compact, nonoverlapping clusters. It works well in practice, is efficient to train, and can easily be used to clustering other points in the future.
Mixture model A probability model used in clustering. Every cluster is modeled as a probability distribution that its points are drawn from. This makes it plausible for a point in your dataset to be in the gray area between two clusters, where it could plausibly have come from either.
Mutual information A measure of "correlation" between two probability distributions that can be used as a measure of how well a set of clusters corresponds to known ground-truth clusters
Principal component analysis A dimensionality reduction technique where you express your input data points, to a good approximation, as linear combinations of several "principal component" vectors. Examining the components themselves can give insights into the dataset.
PCA Popular shorthand for Principal Component Analysis
Rand index A measure of how well a set of clusters corresponds to known ground-truth clusters.
Silhouette score A measure of how well compact and distinct from each other a collection of clusters is.

11

Regression

Regression is similar to classification: you have a number of input features, and you want to predict an output feature. In classification, this output feature is either binary or categorical. With regression, it is a real-valued number.

Typically, regression algorithms fall into two categories:

- Modeling the output as a linear combination of the inputs. There is a ton of elegant math here and principled ways to handle data pathologies.
- Ugly hacks to deal with anything nonlinear.

This chapter will review several of the more popular regression techniques in machine learning, along with some techniques for assessing how well they performed.

I have made the unconventional decision to include fitting a line (or other curves) to two-dimensional data within the chapter on regression. You usually don't see curve fitting in the context of machine learning regression, but they're really the same thing mathematically: you assume some functional form for the output as a function of the inputs (such as $y = m_1 x_1 + m_2 x_2$, where x_i are inputs and m_i are parameters that you set to whatever you want), and then you choose the parameters to line up as well as possible (however, you define "as well as possible") with your training data. The distinction between them is a historical accident; fitting a curve to data was developed long before machine learning and even before computers.

11.1 Example: Predicting Diabetes Progression

The following script uses a dataset describing physiological measurements taken from 442 diabetes patients, with the target variable being an indicator of the progression of their disease. After the script, comes the images it generates.

```
import sklearn.datasets
import pandas as pd
```

The Data Science Handbook, First Edition. Field Cady.

```python
from matplotlib import pyplot as plt
from sklearn.cross_validation import train_test_split
from sklearn.linear_model import\
LinearRegression, Lasso
from sklearn.preprocessing import normalize
from sklearn.metrics import r2_score
diabetes = sklearn.datasets.load_diabetes()
X, Y = normalize(diabetes['data']), diabetes['target']
X_train, X_test, Y_train, Y_test = \
    train_test_split(X, Y, test_size=.8)
linear = LinearRegression()
linear.fit(X_train, Y_train)
preds_linear = linear.predict(X_test)
corr_linear = round(pd.Series(preds_linear).corr(
  pd.Series(Y_test)), 3)
rsquared_linear = r2_score(Y_test, preds_linear)
print("Linear coefficients:")
print(linear.coef_)
plt.scatter(preds_linear, Y_test)
plt.title("Lin. Reg.  Corr=%f Rsq=%f"
  % (corr_linear, rsquared_linear))
plt.xlabel("Predicted")
plt.ylabel("Actual")
# add x=y line for comparison
plt.plot(Y_test, Y_test, 'k--')
plt.show()
lasso = Lasso()
lasso.fit(X_train, Y_train)
preds_lasso = lasso.predict(X_test)
corr_lasso = round(pd.Series(preds_lasso).corr(
  pd.Series(Y_test)), 3)
rsquared_lasso = round(
  r2_score(Y_test, preds_lasso), 3)
print("Lasso coefficients:")
print(lasso.coef_)
plt.scatter(preds_lasso, Y_test)
plt.title("Lasso. Reg.  Corr=%f Rsq=%f"
  % (corr_lasso, rsquared_lasso))
plt.xlabel("Predicted")
plt.ylabel("Actual")
# add x=y line for comparison
plt.plot(Y_test, Y_test, 'k--')
plt.show()
```

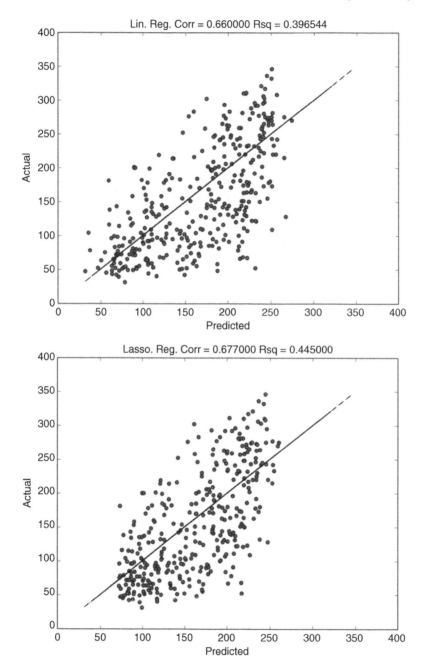

11.2   Least Squares

The simplest example of regression is one that you probably saw in high school: fitting a line to data. You have a collection of x/y pairs, and you try to fit a line to them of the form

$$y = mx + b$$

I remember back in school being encouraged to plot the points out, fit a line to them by eye, trace the line with a ruler, and use that to pull out m and b. On some level, I still think that's the best way to do it, because the human eye can account for outliers and instantly notices data pathologies. Recall Anscombe's quartet, where each of the four datasets has the same line of best fit and the same quality of fit, at least using the standard methods:

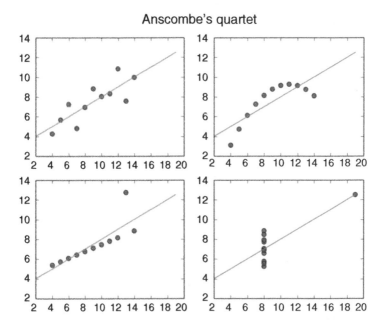

Anscombe's quartet

However, we also need an objective way to pull out a number and one that can be done by a computer without human intervention.

The standard way to fit a line is called least squares. In Python, it can be fit using the Linear Regression class in the example scripts, and the fit coefficients can be found in the following way:

```
>>> import numpy as np
>>> x = np.array([[0.0],[1.0],[2.0]])
```

```
>>> y = np.array([1.0,2.0,2.9])
>>> lm = LinearRegression().fit(x, y)
>>> lm.coef_  # m
array([ 0.95])
>>> lm.intercept_   # b
1.0166666666666671
```

Least squares works by picking the values of *m* and *b* that minimize the "penalty function," which adds up an error term across all of the points:

$$L = \sum_i \left(y_i - (mx_i + b)\right)^2$$

The key thing to understand here is that this penalty function makes least-squares regression extremely sensitive to outliers in the data: three deviations of size 5 will give a penalty of 75, but just a single larger deviation of size 10 will give the larger penalty of size 100. Linear regression will bend the parameters so as to avoid large deviations of even a single point, which makes it unsuitable in situations where a handful of large deviations are to be expected.

An alternative approach that is more suitable to data with outliers is to use the penalty function

$$L = \sum_i \left|y_i - (mx_i + b)\right|$$

where we just take the absolute values of the different error terms and add them. This is called "L1 regression," among other names. Outliers will still have an impact, but it is not as egregious as with least squares. On the other hand, L1 regression penalizes small deviations from expectation more harshly compared to least squares, and it is significantly more complicated to implement computationally.

11.3 Fitting Nonlinear Curves

Fitting a curve to data is a ubiquitous problem not just in data science but in engineering and the sciences in general. Often, there are good a priori reasons that we expect a certain functional form, and extracting the best-fit parameters will tell us something very meaningful about the system we are studying. A few examples that I've seen include the following:

- Exponential decay to some baseline. This is useful for modeling many processes where a system starts in some kind of agitated state and decays to a baseline

$$y = ae^{-bx} + c$$

- Exponential growth

$$y = ae^{bx}$$

- Logistic growth, which is useful in biology for modeling the population density of organisms growing in a constrained environment that can only support so many individuals:

$$y = a\frac{e^{bx}}{c + e^{bx}}$$

- Polynomials of various degrees

$$y = a_0 + a_1 x + a_2 x^2$$

Least squares is the typical approach in all of these cases, where we pick the parameters to as to minimize the penalty function

$$L = \sum_i \left(y_i - f(x_i)\right)^2$$

In Python, the way to do general least-squares fitting is with the curve_fit function, shown in the following code. It takes as its first argument a user-defined function, which takes in x and some number of additional parameters (two parameters in this code), uses those parameters to calculate some function of x, and returns the value. The next arguments are the x values and y values of the data we have. Then curve_fit, through a process of trial-and-error called optimization (which I'll talk about later in the book) tries to find the values of the additional parameters that will minimize the error term for the given x and y values. It returns a tuple of two things: the best-fitted parameters and a matrix that estimates how much they vary.

The following script creates some data of the form $y = 2 + 3x^2$, adds some noise to it, and then uses curve_fit to fit a curve of the form $y = a + bx^2$ to the data.

```
from scipy.optimize import curve_fit
xs = np.array([1.0, 2.0, 3.0, 4.0])
ys = 2.0 + 3.0 *xs*xs + 0.2*np.random.uniform(3)
def calc(x, a, b):
    return a + b*x*x
cf = curve_fit(calc, xs, ys)
best_fit_params = cf[0]
```

When I ran it on my computer, it found $a = 2.33677376$ and $b = 3 - a$ pretty good match.

I should note that, computationally, doing nonlinear fits such as this is extremely slow, and the numerical algorithms can sometimes go horribly awry and give incorrect results. The best way to address this, if you can, is to transform your problem into a linear one by fitting a line to some function of your data (such as the log). If that is not possible, you can often improve the performance by inputting an initial guess as an optional parameter: in curve_fit, that optional argument is called p0.

11.4 Goodness of Fit: R^2 and Correlation

When you are assessing the quality of a fitted curve, there are two questions we want to answer:

- How accurately can we predict values?
- We assumed that the data followed some functional form. Was that even a good assumption?

The standard way to answer the first of the questions is called R^2, pronounced "R squared." R^2 is often described as the fraction of the variance that is accounted for by the model. A value of 1.0 means a perfect match, and a value of 0 means you didn't capture any of the variation. In some cases (there are a few different definitions of R^2 floating around), it can even take on negative values.

If you want to get a bit more detailed, the calculation of R^2 is based on two concepts:

- The total variation:

$$\text{TV} = \sum_i |y_i - \bar{y}|^2$$

where $\bar{y}$ is the average of all the y values in your data.
- The residual variation:

$$\text{RV} = \sum_i |y_i - f(x_i)|^2$$

These allow us to say, in a precise sense, that your fitted model accounts for a certain percentage of the variation in the data. The definition of R^2 is then

$$R^2 = 1 - \left(\frac{\text{RV}}{\text{TV}} \right)$$

and you can see it as the fraction of all variation that is captured by the model. Of course, taking the squares of the residuals isn't necessarily the "right" way to quantify variation, but it is the most standard option.

Despite looking like a square, technically R^2 can be negative if your model is truly abysmal. Having $R^2 = 0$ is what you would see if you just defined your fitted function to return the average of y as a constant value:

$$f(x) = \bar{y}$$

You can think of this as the crudest way to fit a function to data. Do any worse than this, and your R^2 score will go negative.

In the example script at the beginning of the chapter, the relevant lines for R squared were

```
from sklearn.metrics import r2_score
rsquared_linear = r2_score(Y_test, preds_linear)
```

Another way to quantify your goodness-of-fit is to simply take the correlation between your predicted values and the known values in the test data. This has the advantage that you can use Pearson, Spearman, or Kendall correlation, depending on how you want to deal with outliers. On the other hand though, correlation just measures whether your predictions and target values are related; it doesn't measure whether they actually match up.

11.5   Correlation of Residuals

The main ways to measure goodness-of-fit in regression situations are R^2 and correlation between predictions and targets. You typically don't see people asking about whether the functional form being assumed was actually the "correct" form. If you are fitting two-dimensional data though, this question can be addressed as well.

The simplest way to assess the quality of our model form is to plot the known data against a curve of the predicted values. Do they match up? In Anscombe's quartet, for example, it is visually clear that a linear model is the correct way to approach the first dataset, but the wrong way to approach the second one.

A way to quantify this relationship is the so-called correlation of residuals. Intuitively, if our model form is correct, the observed data should be our best-fit formula plus some random noise. In that case, the actual data would be randomly above or below our curve. On the other hand, if our model form was bad, we would expect long stretches where the data was systematically higher or lower than our curve. This suggests that we look across our data points, sorted by x, and calculate the correlation between the consecutive residuals. A correlation near 0 suggests that our model form was good, and any failure of its predictive power comes from true noise in the data, rather than a failure to pick the right functional form.

11.6   Linear Regression

Now let's move on from fitting a curve and into topics that fit more firmly under the "machine learning" umbrella. First up: linear regression.

Linear regression is the same process as fitting a line to data, except that we say

$$y = b + m_1 x_1 + m_2 x_2 + \cdots + m_d x_d$$

where d is the number of input features we have. Most of the previous sections carry over directly to this more general case: we fit the data using least squares, we quantify performance using R^2, and we can also use correlation between predicted and actual values.

The first big difference is that it's no longer practical to plot the predicted curve against the actual data points. What you can do instead is to make a scatterplot between the known test values and the values predicted for those test data points. This allows us to gauge whether our model performs better for larger or smaller values and whether it suffers from major outliers. To illustrate, the aforementioned example script will generate this figure for the linear regression model:

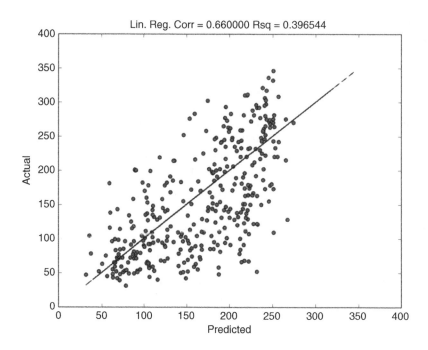

We can see that there is a clear correlation between the predicted and actual numbers, but it is fairly tenuous. In particular, we can see that there are a number of data points where the actual value was substantially below our predictions, that is, their diabetes was significantly less damaging than we would have guessed based on the other measurements. In fact, the fit line as a whole looks slightly more shallow than the data itself. Together, these suggest that there are a number of anomalously low data points, which are pulling our overall predictions lower than perhaps they should be.

The other thing that we can do with linear regression is use it to identify features in the data that are particularly interesting. In the example script, we used the normalize() function to scale all the features so that they had mean 0 and standard deviation 1. This means that, by looking at the relative size of their weights in the linear model, we can get a sense of how related they are to the progression of diabetes. In the example script, I print out the coefficients as the following:

```
>>> print(linear.coef_)
[-28.12698694 -33.32069944  85.46294936  70.47966698
-37.66512686
  20.59488356 -14.6726611  33.10813747  43.68434357
-5.50529361]
```

This suggests that the third and fourth features are particularly interesting, if we want to zero in on and examine their relationship to diabetes more closely.

11.7 LASSO Regression and Feature Selection

Look at the coefficients in the linear regression again. We are able to identify several features as being more promising than the other as targets of further investigation, but the painful truth is that all of the coefficients except that last one are pretty big. There are two problems with this:

- It makes it harder to pinpoint exactly which features are the most interesting.
- There is a very good chance that the data are overfitted. Many of the moderate-sized coefficients could be set so that they balance each other out, yielding a slightly better fit on the training data itself but generalizing very poorly.

The idea of LASSO regression is that we still fit a linear model to the data, but we want to penalize nonzero weights.

A LASSO regression model takes in a parameter called alpha, which indicates how severely nonzero weights should be punished. Setting alpha to 0 will reduce to linear regression. The default value, which was used in the script, is 1.0.

The sample script produces the same scatterplot and performance metrics that were created for linear regression. We can see that the predicted/actual scatterplot hugs the middle line a little more closely, suggesting a better fit. This eyeballing is borne out by the higher R^2 value and correlation. The linear model was indeed overfitting the data.

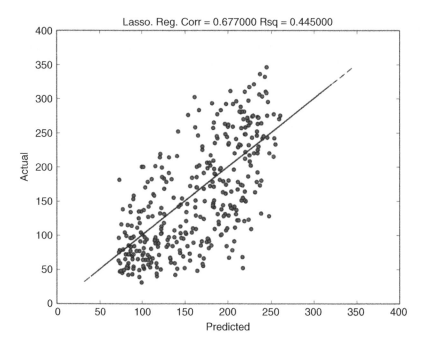

The different between linear and lasso jumps out when we look at the fitted coefficients:

```
>>> print(lasso.coef_)
[ -0.           -11.49747021  73.20707164  37.75257628
  0.            0.
  -10.36895667   3.70576596  24.17976499   0.            ]
```

Four of the six features have weights of precisely 0. Of the remaining features, it is clear that the third is the most relevant to diabetes progression, followed by the fourth and the ninth.

11.8   Further Reading

Bishop, C, *Pattern Recognition and Machine Learning*, 2007, Springer, New York, NY.

Scikit-learn 0.171.1 documentation, http://scikit-learn.org/stable/index.html, viewed 7 August 2016, The Python Software Foundation.

11.9   Glossary

L1 penalty A regression method where we tune our model parameters so as to minimize the sum of the absolute values of the residuals. This method is more robust to outliers than least squares.

Least squares A regression method where we tune our model parameters so as to minimize the sum of the squares of the residuals. This is the most standard best-fit method.

R squared A measure of how well a regression model fits the data. It is the fraction of all the test data's variance that is accounted for by the model.

Residual The difference between the value predicted for a data point and the actual observed value.

12

Data Encodings and File Formats

Coming from a background of academic physics, my first years in data science were one big exercise in discovering new data formats that I probably should have already known about. It was a bit demoralizing at the time, so let me make something clear upfront: people are always dreaming up new data types and formats, and you will forever be playing catch-up on them. However, there are several formats that are common enough you should know them. It seems that every new format that comes out is easily understood as a variation of a previous format, so you'll be on good footing going forward. There are also some broad principles that underlie all formats, and I hope to give you a flavor of them.

First, I will talk about specific file formats that you are likely to encounter as a data scientist. This will include sample code for parsing them, discussions about when they are useful, and some thoughts about the future of data formats.

For the second half of the chapter, I will switch gears to a discussion of how data is laid out in the physical memory of a computer. This will involve peaking under the hood of the computer to look at performance considerations and give you a deeper understanding of the file formats we just discussed. This section will come in handy when you are dealing with particularly gnarly data pathologies or writing code that aims for speed when you are chugging through a dataset.

12.1 Typical File Format Categories

There are many, many different specific file formats out there. However, they fall under several broad categories. This section will go over the most important ones for a data scientist. The list is not exhaustive, and neither are the categories mutually exclusive, but this should give you a broad lay of the land.

The Data Science Handbook, First Edition. Field Cady.
© 2017 John Wiley & Sons, Inc. Published 2017 by John Wiley & Sons, Inc.

12.1.1 Text Files

Most raw data files seen by data scientists are, at least in my experience, text files. This is the most common format for CSV files, JSON, XML and web pages. Pulls from databases, data from the web, and log files generated by machines are all typically text. The advantage of a text file is that it is readable by a human being, meaning that it is very easy to write scripts that generate it or parse it. Text files work best for data with a relatively simple format.

There are limitations though. In particular, text is a notoriously inefficient way to store numbers. The string "938238234232425123" takes up 18 bytes, but the number it represents would be stored in memory as 8 bytes. Not only is this a price to pay for storage, but the number must be converted from text to a new format before a machine can operate on it.

12.1.2 Dense Numerical Arrays

If you are storing large arrays of numbers, it is much more space- and performance-efficient to store them in something such as the native format that computers use for processing numbers. Most image files or sound files consist mostly of dense arrays of numbers, packed adjacent to each other in memory. Many scientific datasets fall into this category too. In my experience, you don't see these datasets as often in data science, but they do come up.

12.1.3 Program-Specific Data Formats

Many computer programs have their own specialized file format. This category would include things such as Excel files, db files, and similar formats. Typically, you will need to look up a tool to open one of these files.

In my experience, opening them often takes a while computationally, since there are often a lot of bells and whistles built into the program that may or may not be present in this particular dataset. This makes it a pain to reparse them every time you rerun your analysis scripts – often, it takes much longer than the actual analysis does. What I typically do is make CSV versions of them right upfront and use those as the input to my analyses.

12.1.4 Compressed or Archived Data

Many data files, when stored in a particular format, take up a lot more space compared to the file in question logically needs; for example, if most lines in a large text file are exactly the same or a dense numerical array consists mostly of 0s. In these cases, we want to compress the large file into a smaller one, so that it can be stored and transferred more easily. A related problem is when we have a large collection of files that we want to condense into a single file for easier management, often called data archiving. There are a variety of ways that we can encode the raw data into these more manageable forms.

There is a lot more to data compression than just reducing the size. A "perfect" algorithm would have the following properties:

- It generally reduces the size of the data, easing storage requirements.
- If it can't compress the data much (or at all), then at least it doesn't balloon it to take up much MORE space.
- You can decompress it quickly. If you do this really well, it might take you less time to load the compressed data compared to the raw data itself, even with the decompression step. This is because decompression in RAM can be fairly quick, but it takes a long time to pull extra data off the disk.
- You can decompress it "one line at a time," rather than loading the entire file. This helps you deal with corrupt data and typically makes decompression go faster since you're operating on less data at a time.
- You can recompress it quickly.

In the real world, there is a wide range of compression algorithms available, which balance these interests in a lot of different ways. Compression becomes especially important in Big Data settings, where datasets are typically large and reloaded from disk every time the code runs.

12.2   CSV Files

CSV files are the workhorse data format for data science. "CSV" usually stands for "comma-separated value," but it really should be "character-separated value" since characters other than commas do get used. Sometimes, you will see ".tsv" if tabs are used or ".psv" if pipes (the "|" character) are used. More often though, in my experience, everything gets called CSV regardless of the delimiter.

CSV files are pretty straightforward conceptually – just a table with rows and columns. There are a few complications you should be aware of though:

- Headers. Sometimes, the first line gives names for all the columns, and sometimes, it gets right into the data.
- Quotes. In many files, the data elements are surrounded in quotes or another character. This is done largely so that commas (or whatever the delimiting character is) can be included in the data fields.
- Nondata rows. In many file formats, the data itself is CSV, but there are a certain number of nondata lines at the beginning of the file. Typically, these encode metadata about the file and need to be stripped out when the file is loaded into a table.
- Comments. Many CSV files will contain human-readable comments, as source code does. Typically, these are denoted by a single character, such as the # in Python.

- Blank lines. They happen.
- Lines with the wrong number of columns. These happen too.

The following Python code shows how to read a basic CSV file into a data frame using Pandas:

```python
import pandas
df = pandas.read_csv("myfile.csv")
)
```

If your CSV file has weird complexities associated with it, then read_csv has a number of optional arguments that let you deal with them. Here is a more complicated call to read_csv:

```python
import pandas
df = pandas.read_csv("myfile.csv",
 sep = "|", # the delimiter. Default is the comma
 header = False,
 quotechar = '"',
 compression = "gzip",
 comment = '#'
)
```

In my work, the optional arguments I use most are sep and header.

12.3 JSON Files

JSON is probably my single favorite data format, for its dirt simplicity and flexibility. It is a way to take hierarchical data structures and serialize them into a plain text format. Every JSON data structure is either of the following:

- An atomic type, such as a number, a string, or a Boolean.
- A JSONObject, which is just a map from strings to JSON data structures. This is similar to Python dictionaries, except that there are keys in the JSONObject.
- An array of JSON data structures. This is similar to a Python list.

Here is an example of some valid JSON, which encodes a JSONObject map with a lot of substructures:

```json
{
"firstName": "John",
"lastName": "Smith",
"isAlive": true,
```

```
"age": 25,
"address": {
 "streetAddress": "21 2nd Street",
 "city": "New York",
 "state": "NY",
 "postalCode": "10021-3100"
},
"children": ["alice","john",{"name":"alice","birth_order":2}],
"spouse": null
}
```

Note a few things about this example:

- The fact that I've made it all pretty with the newlines and indentations is purely to make it easier to read. This could have all been on one long line and any JSON parser would parse it equally well. A lot of programs for viewing JSON will automatically format it in this more legible way.
- The overall object is conceptually similar to a Python dictionary, where the keys are all strings and the values are JSON objects. The overall object could have been an array too though.
- A difference between JSON objects and Python dictionaries is that all the field names have to be strings. In Python, the keys can be any hashable type.
- The fields in the object can be ordered arrays, such as "children." These arrays are analogous to Python lists.
- You can mix and match types in the object, just as in Python.
- You can have Boolean types. Note though that they are declared in lower case.
- There are also numerical types.
- There is a null supported
- You can nest the object arbitrarily deeply.

Parsing JSON is a cinch in Python. You can either "load" a JSON string into a Python object (a dictionary at the highest level, with JSON arrays mapping to Python lists, etc.) or "dump" a Python dictionary into a JSON string. The JSON string can either be a Python string or be stored in a file, in which case you write from/to a file object. The code looks as follows:

```
>>> import json
>>> json_str = """{"name": "Field", "height":6.0}"""
>>> my_obj = json.loads(json_str)
>>> my_obj
{u'name': u'Field', u'height': 6.0}
>>> str_again = json.dumps(my_obj)
```

Historically, JSON was invented as a way to serialize objects from the JavaScript language. Think of the keys in a JSONObject as the names of the members in an object. However, JSON does NOT support notions such as pointers, classes, and functions.

12.4   XML Files

XML is similar to JSON: a text-based format that lets you store hierarchical data in a format that can be read by both humans and machines. However, it's significantly more complicated than JSON – part of the reason that JSON has been eclipsing it as a data transfer standard on the web.

Let's jump in with an example:

```
<GroupOfPeople>
<person gender="male">
<Name>Field Cady</Name>
<Profession>Data Scientist</Profession>
</person>
<person gender="female">
<Name>Ryna</Name>
<Profession>Engineer</Profession>
</person>
</GroupOfPeople>
```

Everything enclosed in angle brackets is called a "tag." Every section of the document is bookended by a matching pairs of tags, which tell what type of section it is. The closing tag contains a slash "/" after the "<". The opening tag can contain other pieces of information about the section – in this case, "gender" is such an attribute. Because you can have whatever tag names or additional attributes you like, XML lends itself to making domain-specific description languages.

XML sections must be fully nested into each other, so something such as the following is invalid:

```
<a><b></a></b>
```

because the "b" section begins in the middle of the "a" section but doesn't end until the "a" is already over. For this reason, it is conventional to think of an XML document as a tree structure. Every nonleaf node in the tree corresponds to a pair of opening/closing tags, of some type and possibly with some attributes, and the leaf nodes are the actual data.

Sometimes, we want the start and end tag of a section to be adjacent to each other. In this case, there is a little bit of syntactic sugar, where you put the closing "/" before the closing angle bracket. So,

```
<foo a="bar"></foo>
```

is equivalent to

```
<foo a="bar"/>
```

A big difference between JSON and XML is that the content in XML is ordered. Every node in the tree has its children in a particular order – the order in which they come in the document. They can be of any types and come in any order, but there is AN order.

Processing XML is a little more finicky than processing JSON, in my experience. This is for two reasons:

- It's easier to refer to a named field in a JSON object than to search through all the children of a XML node and find the one you're looking for.
- XML nodes often have additional attributes, which are handled separately from the node's children.
- This isn't inherent to the data formats, but in practice, JSON tends to be used in small snippets, for smaller applications where the data has regular structure. So, you typically know exactly how to extract the data you're looking for. In contrast, XML is liable to be a massive document with many parts, and you have to sift through the whole thing.

In Python, the XML library offers a variety of ways of processing XML data. The simplest is the ElementTree sublibrary, which gives us direct access to the parse tree of the XML. It is shown in this code example, where we parse XML data into a string object, access and modify the data, and then reencode it back to an XML string:

```
>>> import xml.etree.ElementTree as ET
>>> xml_str = """
<data>
 <country name="Liechtenstein">
   <rank>1</rank>
   <year>2008</year>
   <gdppc>141100</gdppc>
   <neighbor name="Austria" direction="E"/>
   <neighbor name="Switzerland" direction="W"/>
 </country>
 <country name="Singapore">
   <rank>4</rank>
   <year>2011</year>
   <gdppc>59900</gdppc>
   <neighbor name="Malaysia" direction="N"/>
 </country>
```

```
<country name="Panama">
  <rank>68</rank>
  <year>2011</year>
  <gdppc>13600</gdppc>
  <neighbor name="Costa Rica" direction="W"/>
  <neighbor name="Colombia" direction="E"/>
</country>
</data>
"""
>>> root = ET.fromstring(xml_str)
>>> root.tag
'data'
>>> root[0] # gives the zeroth child
<Element 'country' at 0x1092d4410>
>>> root.attrib # dictionary of node's attributes
{}
>>> root.getchildren()
[<Element 'country' at 0x1092d4410>, <Element
'country' at 0x1092d47d0>, <Element 'country' at
0x1092d4910>]
>>> del root[0] # deletes the zeroth child from the tree
>>> modified_xml_str = ET.tostring(root)
```

The "right" way to manage XML data is called the "Document Object Model." It is a little more standardized across programming languages and web browsers, but it is also more complicated to master. The ElementTree is fine for simple applications and capable of doing whatever you need it to do.

12.5 HTML Files

By far the most important variant of XML is HTML, the language for describing pages on the web. Practically speaking, the definition of "valid" HTML is that your web browser will parse it as intended. There are differences between browsers, some intentional and some not, and that's why the same page might look different in Chrome and Internet Explorer. But browsers have largely converged on a standard version of HTML (the most recent official standard is HTML5), and to a first approximation, that standard is a variant of XML. Many web pages could be parsed with an XML parser library.

I mentioned in the last section that XML can be used to create domain-specific languages, each of which is defined by its own set of valid tags and their associated attributes. This is the way HTML works. Some of the more notable tags are given in the following table:

Tag	Meaning	Example
<a>	Hyperlink	Click <a href="www.google.com">here</a> to go to Google
<img>	Image	`<img src="smiley.gif">`
<h1>–<h6>	Headings of text	<h1>The Title</h1>
<div>	Division. It doesn't get rendered but helps to organize the document. Often, the "class" attribute is used to associate the contents of the division with a desired style of text formatting	<div class="main-text">My body of text</div>
<ul> and <li>	Unordered lists (usually rendered as bulleted lists) and list items	Here is a list: <ul> <li>Item One</li> <li>Item Two</li> </ul>

The practical problem with processing HTML data is that, unlike JSON or even XML, HTML documents tend to be extremely messy. They are often individually made, edited by humans, and tweaked until they look "just right." This means that there is almost no regularity in structure from one HTML document to the next, so the tools for processing HTML lean toward combing through the entire document to find what it is you're looking for.

The default HTML tool for Python is the HTMLParser class, which you use by creating a subclass that inherits from it. An HTMLParser works by walking through the document, performing some action each time it hits a start or an end tag or other piece of text. These actions will be user-defined methods on the class, and they work by modifying the parser's internal state. When the parser has walked through the entire document, its internal state can be queried for whatever it is you were looking for. One very important note is that it's up to the user to keep track of things such as how deeply nested you are within the document's sections.

To illustrate, the following code will pull down the HTML for a Wikipedia page, step through its content, and count all hyperlinks that are embedded in the body of the text (i.e., they are within paragraph tags):

```
from HTMLParser import HTMLParser
import urllib

TOPIC = "Dangiwa_Umar"
url = "https://en.wikipedia.org/wiki/%s" % TOPIC
class LinkCountingParser(HTMLParser):
    in_paragraph = False
    link_count = 0
```

```
    def handle_starttag(self, tag, attrs):
        if tag=='p': self.in_paragraph = True
        elif tag=='a' and self.in_paragraph:
            self.link_count += 1
    def handle_endtag(self, tag):
        if tag=='p': self.in_paragraph = False
html = urllib.urlopen(url).read()
parser = LinkCountingParser()
parser.feed(html)
print "there were", parser.link_count, \
    "links in the article"
```

12.6 Tar Files

Tar is the most popular example of an "archive file" format. The idea is to take an entire directory full of data, possibly including nested subdirectories, and combine it all into a single file that you can send in an e-mail, store somewhere, or whatever you want. There are a number of other archive file formats, such as ISO, but in my experience, tar is the most common example.

Tarring a directory doesn't actually compress the data – it just combines the files into one file that takes up about as much space as the data did originally. So in practice, Tar files are almost always then zipped. GZipping in particular is popular. The ".tgz" file extension is used as a shorthand for ".tar.gz", that is, the directory has been put into a Tar file, which was then compressed using the GZIP algorithm.

Tar files are typically opened from the command line, such as the following:

```
$ # This will expand the contents of
$ # my_directory.tar into the local directory
$ tar -xvf my_directory.tar
$ # This command will untar and unzip
$ # a directory with has been tarred and g-zipped
$ tar -zxf file.tar.gz
$ # This command will tar the Homework3 directory
$ # into the file ILoveHomework.tar
$ tar -cf ILoveHomework.tar Homework3
```

12.7 GZip Files

Gzip is the most common compression format that you will see on Unix-like systems such as Mac and Linux. Often, it's used in conjunction with Tar to archive the contents of an entire directory. Encoding data with gzip is comparatively slow, but the format has the following advantages:

- It compresses data super well.
- Data can be decompressed quickly.
- It can also be decompressed one line at a time, in case you only want to operate only on part of the data without decompressing the whole file.

Under the hood, gzip runs on a compression algorithm called DEFLATE. A compressed gzip file is broken into blocks. The first part of each block contains some data about the block, including how the rest of the block is encoded (it will be some type of Huffman code, but you don't need to worry about the details of those). Once the gzip program has parsed this header, it can read the rest of the block 1 byte at a time. This means there is minimal RAM being used up, so all the decompression can go on near the top of the RAM cache and hence proceed at breakneck speed.

The typical commands for gzipping/unzipping from the shell are simple:

```
$ gunzip myfile.txt.gz # creates raw file myfile.txt
$ gzip myfile.txt # compresses the file into myfile.
txt.gz
```

However, you can typically also just double-click on a file – most operating systems can open gzip files natively.

12.8 Zip Files

Zip files are very similar to Gzip files. In fact, they even use the same DEFLATE algorithm under the hood! There are some differences though, such as the fact that ZIP can compress an entire directory rather than just individual files.

Zipping and unzipping files are as easy with ZIP as with GZIP:

```
$ # This puts several files into a single zip file
$ zip filename.zip input1.txt input2.txt resume.doc
pic1.jpg
$ # This will open the zip file and put
$ # all of its contents into the current directory
$ unzip filename.zip
```

Besides their widespread use for archiving files, ZIP files are also used by the Java programming language and its relatives. Compiled Java classes are stored as JAR files, but JAR files are created just by zipping individual Java class files together. JAR is the same format as ZIP, except that you are only combining Java class files rather than arbitrary file types.

12.9 Image Files: Rasterized, Vectorized, and/or Compressed

Image files can be broken down into two broad categories: rasterized and vectorized. Rasterized files break an image down into an array of pixels and encode things such as the brightness or color of each individual pixel. Sometimes, the image file will store the pixel array directly, and other times, it will store some compressed version of the pixel array. Almost all machine-generated data will be rasterized.

Vectorized files, on the other hand, are a mathematical description of what the image should look like, complete with perfect circles, straight lines, and so on. They can be scaled to any size without losing resolution. Vectorized files are more likely to be company logos, animations, and similar things. The most common vectorized image format you're likely to run into is SVG, which is actually just an XML file under the hood (as I mentioned before, XML is great for domain-specific languages!). However, in daily work as a data scientist, you're most likely to encounter rasterized files.

A rasterized image is an array of pixels that, depending on the format, can be combined with metadata and then possibly subjected to some form of compression (sometimes using the DEFLATE algorithm, such as GZIP). There are several considerations that differentiate between the different formats available:

- Lossy versus lossless. Many formats (such as BMP and PNG) encode the pixel array exactly – these are called lossless. But others (such as JPEG) allow you to reduce the size of the file by degrading the resolution of your image.
- Grayscale versus RBG. If images are black-and-white, then you only need one number per pixel. But if you have a colored image, then there needs to be some way to specify the color. Typically, this is done by using RGB encoding, where a pixel is specific by how much red, how much green, and how much blue it contains.
- Transparency. Many images allow pixels to be partly transparent. The "alpha" of a pixel ranges from 0 to 1, with 0 being completely transparent and 1 being completely opaque.

Some of the most important image formats you should be aware of are as follows:

- JPEG. This is probably the single most important one in web traffic, prized for its ability to massively compress an image with almost invisible degradation. It is a lossy compression format, stores RGB colors, and does not allow for transparency.
- PNG. This is maybe the next most ubiquitous format. It is lossless and allows for transparency pixels. Personally, I find the transparent pixels make PNG files super useful when I'm putting together slide decks.

- TIFF. Tiff files are not common on the Internet, but they are a frequent format for storing high-resolution pictures in the context of photography or science. They can be lossy or lossless.

The following Python code will read an image file. It takes care of any decompression or format-specific stuff under the hood and returns the image as a NumPy array of integers. It will be a three-dimensional array, with the first two dimensions corresponding to the normal width and height. The image is read in as RBG by default, and the third dimension of the array indicates whether we are measuring the red, blue, or green content. The integers themselves will range from 0 to 255, since each is encoded with a single byte.

```
from scipy.ndimage import imread
img = imread('mypic.jpg')
```

If you want to read the image as grayscale, you can pass mode = "F" and get a two-dimensional array. If you instead want to include the alpha opacity as a fourth value for each pixel, pass in mode = "RGBA."

12.10   It's All Bytes at the End of the Day

At the lowest level, the data in a computer file is a long array of bits, each of which is set to 0 or 1. That array is broken into 8-bit chunks called bytes. The concept of a byte is both conceptual and physical. On the one hand, we usually break up a file into basic logical units that are composed of bytes, such as having one byte to encode a letter or a number. You could theoretically create a file format where the basic units were of 5 bits or 11 bits long, but the universal convention is to use bytes. At the same time, the physical hardware of the computer is optimized to process data one byte (or a group of several bytes) at a time.

A modern computer's memory is called RAM, for "random access memory." "Random" in this case isn't about probability: it refers to the fact that you can read/modify any part of memory with about the same latency. The memory is physically broken up into bytes for easier processing. The data structures that exist in memory as a program runs are, similarly to raw files, ultimately encoded into bytes. Sometimes, the encodings used for a file and a real-time data structure are identical, and sometimes, they are quite different.

Historically, an atomic type in a programming language was defined to take up a fixed number of bytes. An integer would frequently be allocated 4 bytes, and the integer was encoded in those bytes in binary. Having every integer take up the same amount of space was critical, because the physical layout of the bits doesn't make it clear where one integer (or any other type of variable) ends and another begins. These transitions generally occur on the boundaries between bytes, but that's it. The computer's "native language" doesn't have any notion of integers or any other data type; to the computer, everything is just bytes, so fixed-size variables are critical for keeping track of things.

In modern languages such as Python, there are more variable-size types. For example, a Python string can take up arbitrarily many bytes. However, doing this requires overhead to keep track of where one item ends and another begins, which translates to a substantial performance cost. Modern languages tend to try for fixed-size atomic types whenever possible, but then revert to the less-efficient version when necessary. Software that is intended to run extremely fast, like Python's numerical libraries, almost always strips out the overhead and limits itself to fixed-size types.

12.11   Integers

Integers are about the simplest atomic type to understand. Back in the day when RAM was more expensive, people did all kinds of tricks to try and encode integers using fewer bits, but now things have basically settled out:

- An integer gets a fixed number of bytes. Eight bytes is also typical, if you're using a 64-bit computer.
- The integer is encoded in those bits in binary.
- One of the bits isn't interpreted as a binary digit: it is a flag saying whether the integer is negative. If it's negative, then typically the 0s and 1s are flipped in the rest of the number, for arithmetic efficiency reasons that you don't need to worry about.

This system works seamlessly most of the time, but there is a maximum size of integer that can be handled; 63 bits is only so big. In Python, you can get that upper bound in the following way:

```
>> import sys
>> sys.maxint
9223372036854775807
```

This number is 2 to the power of 63: 63 bits to store the number and one to flag that it's positive. This number is large enough for almost all purposes, but occasionally you need something bigger. Oftentimes, you never even realize that you've ventured into this area! In Python, if you ever declare a variable equal to something larger than sys.maxint, then Python will silently switch over to a different, far less efficient data type called a "long." From a programmer's perspective, a long looks, feels, and acts as an int: the only clear sign that it's something different is a telltale "L" after the number when it's displayed:

```
>>> 3*sys.maxint
2767011611056432742lL
```

The seamless transition is a luxury afforded by using a very high-level language such as Python, and you pay for it in efficiency. It takes overhead to check at every step whether the system needs to switch over to using longs, and if things ever DO switch over the performance hit really cranks up.

12.12 Floats

Floating-point numbers are more complicated than integers, mostly because they are inherently error-prone. A floating number can theoretically have infinitely many decimal places, and the computer can only store finitely many of them. Innocuous operations such as taking a square root, or even dividing by 3, can balloon a previously tame number into infinite-decimal land.

In almost every computer system, a floating-point value is stored as a pair of two numbers, typically a pair of the integers as discussed in the previous section:

- One integer stores the digits in the binary representation of the number.
- The other stores the location of the decimal point in the number.

The overwhelming advantage to this way of doing things is that it lets us represent both very large and very small numbers with the same degree of accuracy; roundoff error will corrupt the number 1 billion about the same percentage that it hurts the number 1 billionth. Other floating-point schemes were tried out back in the day, but they are now in the dustbin of history.

As a data scientist, you don't need to worry too much about roundoff error; good partial workarounds have been baked into most numerical algorithms, and the fact that RAM is so cheap now means that we usually carry around many more decimal places than are ever necessary. However, roundoff issues can show up in subtle ways, as shown in this script:

```
>>> x, y = 0.1, 0.2
>>> x + y
0.30000000000000004
```

This is because 0.1 and 0.2 both have infinitely many decimal places when expressed in binary, so Python only stores an approximation to them. The stored value of x is not 0.1; it is the closest number to 0.1 that can be stored as a float. In this case, that number is slightly larger than 0.1, and similarly for 0.2. When you add x and y, these small errors are large enough to add up. If you try to look at the value of x, you will see

```
>>> x
0.1
```

This number is an illusion! Python is rounding x by a tiny bit before displaying it, as a visual courtesy to the user. But the error margin on $x + y$ is large enough that Python will display it instead.

As with large integers, there are computationally very expensive workarounds for the limitations of machine floating points. Usually, these take the form of either storing numbers as arbitrary-length strings or storing the arithmetic expressions that generated the numbers. These exceedingly expensive, but technically exact, expressions are carried through a computation and can later be approximately cast into the normal style of numbers.

Personally I've never used an exact arithmetic system, and I don't expect to. They are mostly useful in theoretical math situations where exact equality is important, and this almost never occurs in real-world work.

12.13 Text Data

The previous two subsections were kind of academic: you generally don't need to worry about how machines represent numbers in your daily work. This is not the case with strings though: there are several different ways that strings are stored, which have very different tradeoffs, and you must keep an eye toward them. In fact, as I write this, I'm grappling with some nagging string-type issues in my own work. The code isn't correctly converting between two different string implementations, and it's irritating because I thought I fixed the damn thing a while ago. I'm doing this in Python; using a high-level language does not necessarily shield you from string implementation issues.

The granddaddy string format is called ASCII (pronounced "ass", "key"). It is dirt simple, is super efficient, and has stood the test of time. The problem is that it's set in stone, and it's limited. Anything you can type with a standard American-style keyboard can be encoded into ASCII, so you can do a lot with it. But in many modern applications that's not enough. There are Chinese characters, and German letters with an umlaut on top. There are emoticons. There might even be additional types of text that get invented later – just look at the rise of the emoticon!

In ASCII, every character is encoded in a single byte, sometimes called a "char." This gives us an interesting phenomenon: there is a mapping between

ASCII characters and short integers, since they are encoded by the same byte. It's not one-to-one, because ASCII only specifies characters for numbers up to 127, but a byte can encode up to 255 (some bytes are not valid ASCII, but they are still perfectly fine encodings of integers). Quite rationally, the capital "A" is the number 65, "B" is 66, and so on. The lowercase numbers are later, with 97 for "a," 98 for "b," 99 for "c," and so on. Python lets you convert between these using the functions chr() and ord() (for "ordinal"):

```
>>> chr(65)
"A"
>>> ord("A")
65
```

ASCII also includes the various special characters that you can type with a keyboard. Tab is 9, and newline is 10. "@" is 64. The digits "0" through "9" are 48 to 57.

You might think of ASCII as the "establishment" string format that things use if they have to be extremely fault tolerant. Python code is supposed to be stored as ASCII – the Python interpreter will throw an error if you point it at a file that is not ASCII formatted. Operating systems use it. Plain text files are typically ASCII whenever possible. Python string objects are stored in RAM as ASCII.

This might be a good time to revisit the way we declare strings in Python – this paragraph is optional but interesting. Recall that for the most part we just put the contents of the string in quotation marks, and type whatever we want, as in:

```
>>> my_string = "abc123"
```

But some characters, such as tabs and newlines, can't always be directly typed. In this case, we use the slash character "\" to encode them out of things that we CAN type. For example:

```
>>> my_tab = "\t" # this is a one-character string
>>> my_newline = "\n" # this is too
```

Adding the slash before a character in order to encode something is called "escaping" the character. Now I'll give you the keys to the kingdom: if you want super fine-grained control, you can escape "x" to tell the computer exactly which ASCII bytes should be in a string. If I declare a string such as

```
>>> fancy_string = '\xAA'
```

then the two characters "AA" will be interpreted as the hexadecimal number of the ASCII byte you want. Hexagesimal is a slightly archaic, base-16 way to write numbers, whose 16 digits are 0, 1, 2, ..., 9, A, B, ..., E, F. Writing "\t" is just a nicer way of writing "\x09," and "\n" is the same thing as "\x0A" (0A in hexagesimal is 10). In fact, this is more powerful than ASCII, because technically ASCII numbers only go up through 127, whereas hexagesimal notation lets you put in bytes up to 255, that is, any possible byte. Personally, the only time I use hexagesimal notation is when I'm deliberately creating perverse strings for purposes of unit testing, but it's there if you want it.

The other big string standard is known as Unicode. Unicode is actually a family of encoding standards, all of them aiming to supplement basic ASCII with the massive range of other characters needed today and possibly in the future. The main version of Unicode available is UTF-8, and it is fast becoming the most popular encoding around. In this chapter, UTF-8 will be the one I discuss.

The biggest difference between Unicode and ASCII is that in Unicode there is a variable number of bytes that encode each character. This means that all the performance advantages of fixed-sized elements go out the window, but this is the price you must pay for flexibility. However, UTF-8 is backward compatible with ASCII: a chunk of bytes that are valid ASCII are also valid UTF-8. This works because not all bytes are valid ASCII – the ASCII integers top out at 127, but a byte can go up to 255. So, if you are reading through an array of Unicode and come to a byte that is greater than 127, it signifies that this byte (and possibly the next several) constitutes a non-ASCII character. When you upgrade from ASCII to 2-byte characters, you get pretty much all characters in Western languages. Three bytes will give you East Asia. Four will give you various historical writing systems, mathematical symbols, and emoticons.

Python has native support for Unicode. Declaring a string-type variable to be Unicode rather than a normal string is as simple as putting a "u" outside the parentheses:

```
>>> unicode_str = u"This is unicode"
```

Python strings are more general compared to ASCII or UTF-8. Using the "\x" trick, you can force Python to put an arbitrary collection of bytes into a string object, and they may or may not be valid ASCII/Unicode. If you have a string and you want to convert it into valid ASCII or UTF-8, then you can say

```
>>> # fails if not valid ASCII
>>> as_ascii = my_string.decode('ascii')
>>> # drops non-ASCII characters
>>> as_ascii = my_string.decode('ascii', 'ignore')
>>> # drops non-unicode characters
>>> as_utf8 = my_string.decode('utf8', 'ignore')
```

12.14   Further Reading

1 Murrell, P, *Introduction to Data Technologies*, viewed 8 August 2016, http:// statmath.wu.ac.at/courses/data-analysis/
2 Pilgrim, M, 2004, *Dive into Python: Python from Novice to Pro*, viewed 7 August 2016, http://www.diveintopython.net/

12.15   Glossary

Archiving Combining several files or directories into a single file which can later be expanded out to re-create the original files and directories.

ASCII A text encoding scheme that has one character per byte. It pretty much only covers characters that you are likely to type with a standard American keyboard.

Bit A single piece of data that can be either 0 or 1.

Byte Eight bits. Computer memory is physically grouped into bytes.

Compression Taking a single file and condensing it down into a smaller file. Typically, this involves looking for redundancy in the file's contents and seeing how it can be efficiently encoded.

Unicode A family of text encoding schemes that cover many more characters compared to ASCII (especially alphabets from other languages). A single character may require a variable number of bytes to encode.

UTF-8 The most popular Unicode specification.

13

Big Data

There is a lot of overlap between the terms "data science" and "big data." In practice, there is a close relationship between them, but really they mean separate things. Big Data refers to several trends in data storage and processing, which have posed new challenges, provided new opportunities, and demanded new solutions. Often, these Big Data problems required a level of software engineering expertise that normal statisticians and data analysts weren't able to handle. It also posed a lot of difficult, ill-posed questions such as how best to segment users based on raw click-stream data. This demand is what turned "data scientist" into a new, distinct job title. But modern data scientists tackle problems of any scale and only use Big Data technologies when they're the right tool for the job.

Big Data is also an area where low-level software engineering concerns become especially important for data scientists. It's always important that they think hard about the logic of their code, but performance concerns are a strictly secondary concern. In Big Data though, it's easy to accidentally add several hours to your code's runtime, or even have the code fail several hours in due to a memory error, if you do not keep an eye on what's going on inside the computer.

This chapter will start with an overview of two pieces of Big Data software that are particularly important: the Hadoop file system, which stores data on clusters, and the Spark cluster computing framework, which can process that data. I will then move on to some of the fundamental concepts that underlie Big Data frameworks and cluster computing in general, including the famed MapReduce (MR) programming paradigm.

13.1   What Is Big Data?

"Big Data," as the term is used today, is a bit of a misnomer. Massive datasets have been around for a long time, and nobody gave them a special name. Even

The Data Science Handbook, First Edition. Field Cady.
© 2017 John Wiley & Sons, Inc. Published 2017 by John Wiley & Sons, Inc.

today, the largest datasets around are generally well outside of the "big data" sphere. They are generated from scientific experiments, especially particle accelerators, and processed on custom-made architectures of software and hardware.

Instead, Big Data refers to several related trends in datasets (one of which is size) and to the technologies for processing them. The datasets tend to have two properties:

1) They are, as the name suggests, big. There is no special cutoff for when a dataset is "big." Roughly though, it happens when it is no longer practical to store or process it all on a single computer. Instead, we use a cluster of computers, anywhere from a handful of them up to many thousands. The focus is on making our processing scalable, so that it can be distributed over a cluster of arbitrary size with various parts of the analysis going on in parallel. The nodes in the cluster can communicate, but it is kept to a minimum.

2) The second thing about Big Datasets is that they are often "unstructured." This is a terribly misleading term. It doesn't mean that there is no structure to the data, but rather that the dataset doesn't fit cleanly into a traditional relational database, such as SQL. Prototypical examples would be images, PDFs, HTML documents, Excel files that aren't organized into clean rows and columns, and machine-generated log files. Traditional databases presuppose a very rigid structure to the data they contain, and in exchange, they offer highly optimized performance. In Big Data though, we need the flexibility to process data that comes in any format, and we need to be able to operate on that data in ways that are less predefined. You often pay through the nose for this flexibility when it comes to your software's runtime, since there are very few optimizations that can be prebuilt into the framework.

Big Data requires a few words of caution. The first is that you should be hesitant about using Big Data tools. They're all the rage these days, so many people are jumping on the bandwagon blindly. But Big Data tools are almost always slower, harder to set up, and more finicky than their traditional counterparts. This is partly because they're new technologies that haven't matured yet, but it's also inherent to the problems they're solving: they need to be so flexible to deal with unstructured data, and they need to run on a cluster of computers instead of a stand-alone machine. So if your datasets will always be small enough to process with a single machine, or you only need operations that are supported by SQL, you should consider doing that instead.

The final word of caution is that even if you are using Big Data tools, you should probably still be using traditional technologies in conjunction with them. For example, I very rarely use Big Data to do machine learning or data visualization. Typically, I use Big Data tools to extract the relevant features from my data. The extracted features take up much less space compared to the raw dataset, so I can then put the output onto a normal machine and do the actual analytics using something such as Pandas.

I don't mean to knock Big Data tools. They really are fantastic. It's just that there is so much hype and ignorance surrounding them: like all tools, they are great for some problems and terrible for others. With those disclaimers out of the way, let's dive in.

13.2   Hadoop: The File System and the Processor

The modern field of Big Data largely started when Google published its seminal paper on MapReduce, a cluster computing framework it had created to process massive amounts of web data. After reading the paper, an engineer named Doug Cutting decided to write a free, open-source implementation of the same idea. Google's MR was written in C++, but he decided to do it in Java. Cutting named this new implementation Hadoop, after his daughter's stuffed elephant. Hadoop caught on like wildfire and quickly became almost synonymous with Big Data. Many additional tools were developed that ran on Hadoop clusters or that made it easier to write MR jobs for Hadoop.

There are two parts to Hadoop. The first is the Hadoop Distributed File System (HDFS). It allows you to store data on a cluster of computers without worrying about what data is on which node. Instead, you refer to locations in HDFS just as you would for files in a normal directory system. Under the hood, HDFS takes care of what data is stored on which node, keeping multiple copies of the data in case some node fails and other boilerplate.

The second part of Hadoop is the actual MR framework, which reads in data from HDFS, processes it in parallel, and writes its output to HDFS.

I'm actually not going to say much about the Hadoop MR framework, because ironically it's a bit of a dinosaur these days (shows you how quickly Big Data is evolving!). There is a huge amount of overhead for its MR jobs (most damningly, it always reads its input from disk and writes output to disk, and disk IO is much more time-consuming than just doing things in RAM). Additionally, it does a really lousy job of integrating with more conventional programming languages. The community's focus has shifted toward other tools, which still operate on data in HDFS, most notably Spark, and I'll dwell more on them.

13.3   Using HDFS

Data can be accessed and manipulated in HDFS through a command line interface, which is based on the standard bash shell. The following are the main commands you will use:

Example command	What is does
hadoop fs -ls /an/hdfs/location	Display content of the HDFS directory
hadoop fs -copyFromLocal myfile.txt /an/ hdfs/location	Copy data from the local machine to a location in HDFS
hadoop fs -copyToLocal /hdfs/path some/ local/directory	Copy data from HDFS down to the local computer
hadoop fs -cat /path/in/hdfs	Print the contents of a file in HDFS to the screen
hadoop fs -mv /hdfs/location1 /hdfs/ location2	Move data from one location in HDFS to another
hadoop fs -rm /file/in/hdfs	Delete data in HDFS
hadoop fs -rmr /some/hdfs/directory	Recursively delete a directory in HDFS
hadoop fs -appendToFile localfile.txt /user/ hadoop/hadoopfile	Append local data to a file in HDFS

These commands are pretty spartan, but they do everything you need to.

One of the first things you will notice as you work with HDFS is that many of the files have names such as part-m-00000. This is a near-universal convention among data processing tools in Hadoop. The directory containing those files will be the output of a single job that was distributed across multiple nodes in the cluster. The files themselves will be the outputs of different parts of the job that were going on in parallel, and they are stored on the cluster node that generated them.

Occasionally, it's more convenient to just use HDFS rather than the Big Data analysis tools themselves. This is because most of the standard tools involve tremendous overhead, but the hdfs commands are very fast. When used in conjunction with the normal bash shell, you can often get an edge in performance and convenience over the main tools. For example, if my dataset is relatively small, I might do the following to pull out only the rows in a dataset that contain the word "boston" and save them to a local file:

```
hadoop fs -cat /my/dataset/* | grep boston \
    > rows_containing_boston.csv
```

The downside to doing stuff such as this is that the data will all get pulled to the master node in the cluster and processed there – this isn't a parallel thing.

But if the dataset in hdfs is small enough for this to be feasible, it can be a lot quicker than the standard ways.

13.4 Example PySpark Script

PySpark is the most popular way for Python users to work with Big Data. It operates as a Python shell, but it has a library called PySpark, which lets you plug into the Spark computational framework and parallelize your computations across a cluster. The code reads similarly to normal Python, except that there is a SparkContext object whose methods let you access the Spark framework.

This script, whose content I will explain later, uses parallel computing to calculate the number of times every word appears in a text document.

```python
# Create the SparkContext object
from pyspark import SparkConf, SparkContext
conf = SparkConf()
sc = SparkContext(conf=conf)

# Read file lines and parallelize them
# over the cluster in a Spark RDD
lines = open("myfile.txt ")
lines_rdd = sc.parallelize(lines)

# Remove punctuation, make lines lowercase
def clean_line(s):
    s2 = s.strip().lower()
    s3 = s2.replace(".","").replace(",","")
    return s3

lines_clean = lines_rdd.map(clean_line)

# Break each line into words
words_rdd = lines_clean.flatmap(lambda l: l.split())

# Count words
def merge_counts(count1, count2):
    return count1 + count2

words_w_1 = words_rdd.map(lambda w: (w, 1))
counts = words_w_1.reduceByKey(merge_counts)

# Collect counts and display
for word, count in counts.collect():
    print "%s: %i " % (word, count)
```

If Spark is installed on your computer and you are in the Spark home directory, you can run this script on the cluster with the following command:

```
bin/spark-submit --master yarn-client myfile.py
```

Alternatively, you can run the same computation on just a single machine with the following command:

```
bin/spark-submit --master local myfile.py
```

13.5 Spark Overview

Spark is the leading Big Data processing technology these days in the Hadoop ecosystem, having largely replaced traditional hadoop MR. It is usually more efficient, especially if you are chaining several operations together, and it's tremendously easier to use. From a user's perspective, Spark is just a library that you import when you are using either Python or Scala. Spark is written in Scala and runs faster when you call it from Scala, but this chapter will introduce the Python API, which is called PySpark. The example script at the beginning of this chapter was all PySpark. The Spark API itself (names of functions, variables, etc.) is almost identical between the Scala version and the Python version.

The central data abstraction in PySpark is a "resilient distributed dataset" (RDD), which is just a collection of Python objects. These objects are distributed across different nodes in the cluster, and generally, you don't need to worry about which ones are on which nodes. They can be strings, dictionaries, integers – more or less whatever you want. An RDD is immutable, so its contents cannot be changed directly, but it has many methods that return new RDDs. For instance, in the aforementioned example script, we made liberal use of the "map" method. If you have an RDD called X and a function called f, then X.map(f) will apply f to every element of X and return the results as a new RDD.

RDDs come in two types: keyed and unkeyed. Unkeyed RDDs support operations such as map(), which operate on each element of the RDD independently. Often though, we want more complex operations, such as grouping all elements that meet some criteria or joining two different RDDs. These operations require coordination between different elements of an RDD, and for these operations, you need a keyed RDD.

If you have an RDD that consists of two-element tuples, the first element is considered the "key" and the second element the "value." We created a keyed RDD and processed it in the aforementioned script with the following lines:

```
words_w_1 = words_rdd.map(lambda w: (w, 1))
counts = words_w_1.reduceByKey(merge_counts)
```

Here words_w_1 will be a keyed RDD, where the keys are the words and the values are all 1. Every occurrence of a word in the dataset will give rise to a different element in words_w_1. The next line uses the reduceByKey method to group all values that share a key together and then condense them down to a single aggregate value.

I should note that the keyed and unkeyed RDDs are not separate classes in the PySpark implementation. It's just that certain operations you can call (such as reduceByKey) will assume that the RDD is structured as key–value pairs, and it will fail at runtime if that is not the case.

Besides RDDs, the other key abstraction the user has to be aware of is the SparkContext class, which interfaces with the Spark cluster and is the entry point for Spark operations. Conventionally, the SparkContext in an application will be called sc.

Generally, PySpark operations come in two types:

- Calling methods on the SparkContext, which create an RDD. In the example script, we used parallelize() to move data from local space into the cluster as an RDD. There are other methods that will create RDDs from data that is already distributed, by reading it out of HDFS or another storage medium.
- Calling methods on RDDs, which either return new RDDs or produce output of some kind.

Most operations in Spark are what's called "lazy." When you type

```
lines_clean = lines_rdd.map(clean_line)
```

no actual computation gets done. Instead, Spark will just keep track of how the RDD lines_clean is defined. Similarly, lines_rdd quite possibly doesn't exist either and is only implicitly defined in terms of some upstream process. As the script runs, spark is piling up a large dependency structure of RDDs defined in terms of each other, but never actually creating them. Eventually, you will call an operation that produces some output, such as saving an RDD into HDFS or pulling it down into local Python data structures. At that point, the dominos start falling, and all of the RDDs that you have previously defined will get created and fed into each other, eventually resulting in the final side effect. By default, an RDD exists only long enough for its contents to be fed into the next stage of processing. If an RDD that you define is never actually needed, then it will never be brought into being.

The problem with lazy evaluation is that sometimes we want to reuse an RDD for a variety of different processes. This brings us to one of the most important aspects of Spark that differentiates it from traditional Hadoop MR: Spark can cache an RDD in the RAM of the cluster nodes, so that it can be reused as much as you want. By default, an RDD is an ephemeral data structure that only exists long enough for its contents to be passed into the next stage of processing, but a cached RDD can be experimented with in real time. To cache an RDD in memory, you just call the cache() method on it. This method will not actually create the RDD, but it will ensure that the first time the RDD gets created, it is persisted in RAM.

There is one other problem with lazy evaluation. Say again that we write the line

```
lines_clean = lines_rdd.map(clean_line)
```

But imagine that the clean_line function will fail for some value in lines_rdd. We will not know this at the time: the error will only arise later in the script, when lines_clean is finally forced to be created. If you are debugging a script, a tool that I use is to call the count() method on each RDD as soon as it is declared. The count() method counts the elements in the RDD, which forces the whole RDD to be created, and will raise an error if there are any problems. The count() operation is expensive, and you should certainly not include those steps in code that gets run on a regular basis, but it's a great debugging tool.

13.6   Spark Operations

This section will give you a rundown of the main methods that you will call on the SparkContext object and on RDDs. Together, these methods are everything you will do in a PySpark script that isn't pure Python.

The SparkContext object has the following methods:

- sc.parallelize(my_list): Takes in a list of Python objects and distributes them across the cluster to create an RDD.
- sc.textFile("/some/place/in/hdfs"): Takes in the location of text files in HDFS and returns an RDD containing the lines of text.
- sc.pickleFile("/some/place/in/hdfs"): Takes a location in HDFS that stores Python objects that have been serialized using the pickle library. Deserializes the Python objects and returns them as an RDD. This is a really useful method.
- addFile("myfile.txt"): Copies myfile.txt from the local machine to every node in the cluster, so that they can all use it in their operations.
- addPyFile("mylib.py"): Copies mylib.py from the local machine to every node in the cluster, so that it can be imported as a library and used by any node in the cluster.

The main methods you will use on an RDD are as follows:

- rdd.map(func): Applies func to every element in the RDD and returns that results as an RDD.
- rdd.filter(func): Returns an RDD containing only those elements x of rdd for which func(x) evaluates to True.
- rdd.flatMap(func): Applies func to every element in the RDD. func(x) doesn't return just a single element of the new RDD: it returns a list of new elements,

so that one element in the original RDD can turn into many in the new one. Or, an element in the original RDD might result in an empty list and hence no elements of the output RDD.

- rdd.take(5): Computes five elements of RDD and returns them as a Python list. Very useful when debugging, since it only computes those five elements.
- rdd.collect(): Returns a Python list containing all the elements of the RDD. Make sure you only call this if the RDD is small enough that it will fit into the memory of a single computer.
- rdd.saveAsTextFile("/some/place/in/hdfs"): Saves an RDD in HDFS as a text file. Useful for an RDD of strings.
- rdd.saveAsPickleFile("/some/place/in/hdfs"): Serializes every object in pickle format and stores them in hdfs. Useful for RDDs of complex Python objects, such as dictionaries and lists.
- rdd.distinct(): Filters out all duplicates.
- rdd1.union(rdd2): Combines elements of rdd1 and rdd2 into a single RDD.
- rdd.cache(): Whenever RDD is actually created, it will be cached in RAM so that it doesn't have to be re-created later.
- rdd.keyBy(func): This is a simple wrapper for making keyed RDDs, since it is such a common use case. This is equivalent to rdd.map(lambda x: (func(x), x)).
- rdd1.join(rdd2): This works on two keyed RDDs. If (k, v1) is in rdd1 and (k, v2) is in rdd2, then (k, (v1, v2)) will be in the output RDD.
- rdd.reduceByKey(func): For every unique key in the keyed rdd, this collects all of its associated values and aggregates them together using func.
- rdd.groupByKey(func): func will be passed a tuple of two things – a key and an iterable object that will give it all of the values in RDD that share that key. Its output will be an element in the resulting RDD.

13.7 Two Ways to Run PySpark

PySpark can be run either by submitting a stand-alone Python script or by opening up an interpreted session where you can enter your Python commands one at a time. In the previous example, we ran our script by saying

```
bin/spark-submit --master yarn-client myfile.py
```

The spark-submit command is what we use for stand-alone scripts. If instead we wanted to open up an interpreter, we would say

```
bin/pyspark --master yarn-client
```

This would open up a normal-looking Python terminal, from which we could import the PySpark libraries.

From the perspective of writing code, the key difference between a stand-alone script and an interpreter session is that in the script we had to explicitly create the SparkContext object, which we called sc. It was done with the following lines:

```
from pyspark import SparkConf, SparkContext
conf = SparkConf()
sc = SparkContext(conf=conf)
```

If you open up an interpreter though, it will automatically contain the SparkContext object and call it sc. No need to create it manually.

The reason for this difference is that stand-alone scripts often need to set a lot of configuration parameters so that somebody who didn't write them can still run them reliably. Calling various methods on the SparkConf object sets those configurations. The assumption is that if you open an interpreter directly, then you will set the configurations yourself from the command line.

13.8 Configuring Spark

Clusters are finicky things. You need to make sure that every node has the data it needs, the files it relies on, no node gets overloaded, and so on. You need to make sure that you are using the right amount of parallelism, because it's easy to make your code slower by having it be too parallel. Finally, multiple people usually share a cluster, so the stakes are much higher if you hog resources or crash it (which I have done – trust me, people get irate when the whole cluster dies). All of this means that you need to have an eye toward how your job is configured. This section will give you the most crucial parts.

All of the configurations can be set from the command line. The ones you are most likely to have to worry about are the following:

- Name: A human-readable name to give your process. This doesn't affect the running, but it will show up in the cluster monitoring software so that your sys admin can see what resources you're taking up.
- Master: This identifies the "master" process, which deals with parallelizing your job (or running it in local mode). Usually, "yarn-client" will send your job to the cluster for parallel processing, while "local" will run it locally. There are other masters available sometimes, but local and yarn are the most common. Perhaps surprisingly, the default master is local rather than yarn; you have to explicitly tell PySpark to run in parallel if you want parallelism.
- py-files: A comma-separate list of any Python library files that need to be copied to other nodes in the cluster. This is necessary if you want to use that library's functionality in your PySpark methods, because under the hood, each node in the cluster will need to import the library independently.

- Files: A comma-separated list of any additional files that should be put in the working directory on each node. This might include configuration files specific to your task that your distributed functionality depends on.
- Num-executors: The number of executor processes to spawn in the cluster. They will typically be on separate nodes. The default is 2.
- Executor-cores: The number of CPU cores each executor process should take up. The default is 1.

An example of how this might look setting parameters from the command line is as follows:

```
bin/pyspark \
--name my_pyspark_process \
--master yarn-client \
--py-files mylibrary.py,otherlibrary.py \
--files myfile.txt,otherfile.txt \
--num-executors 5 \
--executor-cores 2
```

If instead you want to set them inside a stand-alone script, it will be as follows:

```
from pyspark import SparkConf, SparkContext
conf = SparkConf()
conf.setMaster("yarn-client")
conf.setAppName("my_pyspark_process")
conf.set("spark.num.executors", 5)
conf.set("spark.executor.cores", 2)
sc = SparkContext(conf=conf)
sc.addPyFile("mylibrary.py")
sc.addPyFile("otherlibrary.py")
sc.addFile("myfile.txt")
sc.addFile("otherfile.txt")
```

13.9 Under the Hood

In my mind, PySpark makes a lot more sense when you understand just a little bit about what's going on under the hood. Here are some of the main points:

- When you use the "pyspark" command, it will actually run the "python" command on your computer and just make sure that it links to the appropriate spark libraries (and that the SparkContext object is already in the namespace, if you're running in interactive mode). This means that any Python libraries you have installed on your main node are available to you within your PySpark script.

- When running in cluster mode, the Spark framework cannot run Python code directly. Instead, it will kick off a separate Python process on each node and run it as a separate subprocess. If your code needs libraries or additional files that are not present on a node, then the process that is on that node will fail. This is the reason you must pay attention to what files get shipped around.

- Whenever possible, a segment of your data is confined to a single Python process. If you call map(), flatMap(), or a variety of other PySpark operations, each node will operate on its own data. This avoids sending data over the network, and it also means that everything can stay at Python objects.

- It is very computationally expensive to have the nodes shift data around between them. Not only do we have to actually move the data around, but also the Python objects must be serialized into a string-like format before we send them over the wire and then rehydrated on the other end.

- Operations such as groupByKey() will require serializing data and moving it between nodes. This step in the process is called a "shuffle." The Python processes are not involved in the shuffle. They just serialize the data and then hand it off to the Spark framework.

13.10 Spark Tips and Gotchas

Here are a few parting tips for using Spark, which I have learned from experience and/or hard lessons:

1) RDDs of dictionaries make both code and data much more understandable. If you're working with CSV data, always convert it to dictionaries as your first step. Yeah, it takes up more space because every dictionary has copies of the keys, but it's worth it.
2) Store things in pickle files rather than in text files if you're likely to operate on them later. It's just so much more convenient.
3) Use take() while debugging your scripts to see what format your data is in (RDD of dictionaries? Of tuples? etc.).
4) Running count() on a RDD is a great way to force it to be created, which will bring any runtime errors to the surface sooner rather than later.
5) Do most of your basic debugging in local mode rather than in distributed mode, since it goes much faster if your dataset is small enough. Plus, you reduce the chances that something will fail because of bad cluster configuration.
6) If things work fine in local mode but you're getting weird errors in distributed mode, make sure that you're shipping the necessary files across the cluster.

7) If you're using the --files option from the command line to distribute files across the cluster, make sure that the list is separated by commas rather than colons. I lost two days of my life to that one...

Now that we have seen PySpark in action, let's step back and consider some of what's going on here in the abstract.

13.11   The MapReduce Paradigm

MapReduce is the most popular programming paradigm for Big Data technologies. It makes programmers write their code in a way that can be easily parallelized across a cluster of arbitrary size or an arbitrarily large dataset. Some variant of MR underlies many of the major Big Data tools, including Spark, and probably will for the foreseeable future, so it's very important to understand it.

An MR job takes a dataset, such as a Spark RDD, as input. There are then two stages to the job:

- Mapping. Every element of the dataset is mapped, by some function, to a collection of key–value pairs. In PySpark, you can do this with the flatMap method.
- Reducing. For every unique key, a "reduce" process is kicked off. It is fed all of its associated values one at a time, in no particular order, and eventually, it produces some outputs. You can implement this using the reduceByKey method in PySpark.

And that's all there is to it: the programmer writes the code for the mapper function, and they write the code for the reducer, and that's it. No worrying about the size of the cluster, what data is where, and so on.

In the example script I gave, Spark will end up optimizing the code into a single MR job. Here is the code rewritten so as to make it explicit:

```
def mapper(line):
    l2 = l.strip().lower()
    l3  = l2.replace(".","").replace(",","")
    words = l3.split()
    return [(w, 1) for w in words]

def reducer_func(count1, count2):
    return count1 + count2

lines = open("myfile.txt")
lines_rdd = sc.parallelize(lines)
```

```
map_stage_out = lines_rdd.flatMap(mapper)
reduce_stage_out = \
    map_stage_out.reduceByKey(reducer_func)
```

What happens under the hood in an MR job is the following:

- The input dataset starts off being distributed across several nodes in the cluster.
- Each of these nodes will, in parallel, apply the mapping function to all of its pieces of data to get key–value pairs.
- Each node will use the reducer to condense all of its key–value pairs for a particular word into just a single one, representing how often that word occurred in the node's data. Again, this happens completely in parallel.
- For every distinct key that is identified in the cluster, a node in the cluster is chosen to host the reduce process.
- Every node will forward each of its partial counts to the appropriate reducer. This movement of data between nodes is often the slowest stage of the whole MR job – even slower than the actual processing.
- Every reduce process runs in parallel on all of its associated values, calculating the final word counts.

The overall workflow is displayed in the following diagram:

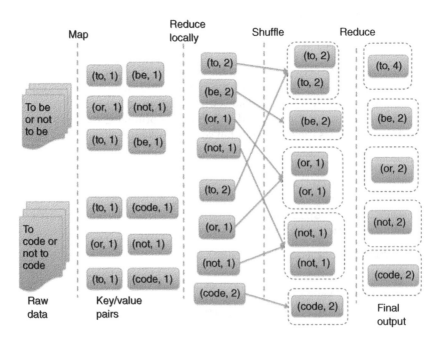

There is one thing I have done here that breaks from classical MR. I have said that each node uses the reducer to condense all of its key–value pairs for a particular word into just one. That stage is technically a performance optimization called a "combiner." I was only able to use a combiner because my reducer was just doing addition, and it doesn't matter what order you add things up in. In the most general case, those mapper outputs are not condensed – they are all sent to whichever node is doing the reducing for that word. This puts a massive strain on the bandwidth between the clusters, so you want to use combiners whenever possible.

13.12 Performance Considerations

There are several guidelines applicable to any MR framework, including Spark:

- If you are going to filter data out, do it as early as possible. This reduces network bandwidth.
- The job only finishes when the last reduce process is done, so try to avoid a situation where one reducer is handling most of the key–value pairs.
- If possible, more reducers means each one has to handle fewer key–value pairs.

In traditional coding, the name of the game in performance optimization is to reduce the number of steps your code takes. This is usually a secondary concern in MR. The biggest concern instead becomes the time it takes to move data from node to node across the network. And the number of steps your code takes doesn't matter so much – instead, it's how many steps your worst node takes.

There is one other specific optimization with Spark in particular that I should mention, which doesn't come up all that often but can be a huge deal when it does. Sometimes, reduceByKey is the wrong method to use. In particular, it is very inefficient when your aggregated values are large, mutable data structure.

Take this dummy code, for example, which takes all occurrences of a word and puts them into a big list:

```
def mapper(line):
    return [(w, [w]) for w in line.split()]
def red_func(lst1, lst2):
    return lst1 + lst2

result = lines.flatMap(mapper).reduceByKey(red_func)
```

As I've written it, every time red_func is called, it is given two potentially very long lists. It will then create a new list in memory (which takes quite a bit of time) and then delete the original lists. This is horribly abusive to the memory, and I've seen jobs die because of it.

Intuitively, what you want to do is keep a big list and just append all the words to it, one at a time, rather than constantly creating new lists. That can be accomplished with the aggregateByKey function, which is a little more complicated to use compared to reduceByKey, but much more efficient if you use it correctly. Example code is here:

```
def update_agg(agg_list, new_word):
    agg_list.append(new_word)
    return agg_list   # same list!

def merge_agg_lists(agg_list1, agg_list2):
    return agg_list1 + agg_list2

def reducer(l1, l2):
    return l1 + l2

result = lines.flatMap(mapper).aggregateByKey(
    [], update_agg, merge_agg_lists)
```

In this case, each node in the cluster will start off with the empty list as its aggregate for a particular word. Then it will feed that aggregate, along with each instance of the word, into update_agg. Then update_agg will append the new value to the list, rather than creating a new one, and return the updated list as a result. The function mergE_agg_lists still operates the original way, but it is only called a few times to merge the outputs of the different nodes.

13.13   Further Reading

1 *Spark Programming Guide*, viewed 8 August 2016, http://spark.apache.org/docs/latest/programming-guide.html
2 Dean, J & Ghemawat, S, MapReduce: Simplified Data Processing on Large Clusters, Paper presented at: Sixth Symposium on Operating System Design and Implementation, 2014, San Francisco, CA

13.14   Glossary

Big data A movement in the analytics and software community that focuses on large, unstructured datasets and how they can be analyzed on clusters of computers.
Combiner A performance optimization in MapReduce frameworks where each node partially completes the reduce process on its own key–value pairs.

Cluster A collection of computers that can be programmed to coordinate a single computation across them.

Hadoop A cluster storage and computing framework that has become the main piece of many Big Data ecosystems.

Key–value pair A tuple with two elements. The second element is the "value," which is usually a piece of data. The first element is a "key," which is usually a label indicating some category the tuple falls into.

Map An operation where you take a collection of data structures and apply the same function to each of them. The outputs of the functions are, collectively, the output of the process.

MapReduce The most prominent paradigm for programming a cluster of computers in a completely parallel way.

Node A single computer in a cluster.

PySpark The Python interface to the Spark cluster computing framework.

RDD Short for Resilient Distributed Dataset.

Reduce An operation where a stream of values are processed one at a time, updating an aggregate with each value. After the last value, the aggregate is returned as the result of the process.

Resilient distributed dataset The abstraction in Spark, an immutable collection of data objects that are distributed across a cluster.

Spark The leading cluster computing framework.

14

Databases

Databases play an important role in data science, but data scientists coming from backgrounds other than programming are often woefully ignorant of them. That was certainly my own experience! In fact, I didn't even really appreciate what the role of a "database" was and why you would use one as opposed to just files of data organized in a directory structure.

A database is ultimately just a framework for storing and efficiently accessing data. A prototypical database is a burly server, which holds more data than would fit into a normal computer, stores it in a way that it can be quickly accessed (this usually involves a ton of under-the-hood optimizations that the database user is blissfully ignorant of), and is standing at the ready to field requests from other computers to access or modify the data. The main advantage of a database relative to raw files is performance, especially if you are running a time-sensitive service (such as a web page). Databases also handle other overhead, such as keeping multiple copies of data synced up and moving data between different storage media.

On the smaller end, many pieces of single-computer software will run a database under the hood, using it as an efficient way to store one program's data. On the larger end, many databases span hundreds of different servers, with complicated protocols for syncing with each other, and users access them over the Internet or a local network. The idea of a single database in this case is really an abstraction that lets the user ignore which of the physical servers they are actually communicating with.

Strictly speaking, a "database" refers to the data itself and its organization, while "database management system" (DBMS) is the software framework that provides access to that data. In practice, these terms are often used interchangeably, and I will be pretty casual about the distinction in this book.

There are many, many ways that databases can be accessed. In production systems, it is often through programmatic APIs that are called through whatever language you are writing your code in. However, most databases also have their own command-line-based shells. This chapter will focus on that use case.

The Data Science Handbook, First Edition. Field Cady.
© 2017 John Wiley & Sons, Inc. Published 2017 by John Wiley & Sons, Inc.

By far the most important family of databases is the SQL family, which supports the relational database (RDB) model (described in the next section). SQL-like databases have been around for a long time. They are generally wickedly fast and support very extensive processing of the data, but in exchange for this power, they are extremely rigid as to the type and formats of the data you can put in them. In recent years, the so-called NoSQL databases are often more flexible in the types of data they will store, but they do not offer the same computational power.

14.1   Relational Databases and MySQL®

In an RDB, a dataset is represented by a table with rows (often unordered) and columns. Each column has a specific data type associated with it, such as an integer, a time stamp, or a string (typically with a known maximum length. They are called VARCHARs in SQL lingo. An RDB has a "query language" associated with it, which lets users specify which data should be selected and any preprocessing/aggregation that should be done before it is returned. The database is structured so that those queries can be answered extremely efficiently.

The SQL family is a large class of relational databases, which have nearly identical interfaces. Of these, MySQL is the most popular open-source version and the one you're most likely to encounter in data science. This section will be based on MySQL, but know that almost everything translates over to the rest of the SQL family. In fact, it will apply well outside of that even; SQL syntax is ubiquitous, and many data processing languages borrow heavily from it.

14.1.1   Basic Queries and Grouping

The data in a MySQL server consists of a collection of tables, whose columns are of known types. The tables are organized into "databases." A database is just a namespace for tables that keeps them more organized; you can switch between namespaces easily or combine tables from several namespaces in a single analysis.

A simple MySQL query will illustrate some of the core syntax:

```
USE my_database;
SELECT name, age
FROM people
WHERE state='WA';
```

The first line is saying that we will be referring to tables in a database called my_database. Next, it is assumed that you have a table called "people" within my_database, with columns name, age, and state (and possibly other columns). This query will give you the name and age of all people living in Washington state. If we had instead said "SELECT *", it would have been shorthand for selecting all of the columns in the table. This selection of rows and columns is the most basic functionality of MySQL.

It is also possible to omit the USE statement and put the name of the database explicitly in your query:

```
SELECT name, age
FROM my_database.people
WHERE state='WA';
```

Aside from just selecting columns, it is also possible to apply operations to the columns before they are returned. MySQL has a wide range of built-in functions for operating on the data fields that can be used in both the SELECT clause and the WHERE clause. For example, the following query will get people's first names and whether or not they are a senior citizen:

```
SELECT SUBSTRING(name,0,LOCATE(name,' ')), (age >= 65)
FROM people
WHERE state='WA';
```

Note the somewhat clunky syntax for getting somebody's first name. The term LOCATE(name,' ') will give us the index of the space in the name, that is, where the first name ends. Then, SUBSTRING(name,0,LOCATE (name,' ')) gives us the name up to that point, that is, the first name. In Python, it would have made more sense to split the string on whitespace and then take the first parts. But doing so calls into existence a list, which can be an arbitrarily complex data structure. This is anathema in performance-focused MySQL! MySQL's functions generally don't permit complex data types such as lists; they limit themselves to functions that can be made blindingly efficient when performed at massive scale. This forces us to extract the first name in this roundabout way.

The following table summarizes a few of the more useful functions:

Function name	Description
ABS	Absolute value
COALESCE	Take the first non-null value of the functions arguments (very useful after joins)
CONCAT	Concatenate several strings
CONVERT_TZ	Convert from one time zone to another
DATE	Extract the date from a datetime expression

Function name	Description
DAYOFMONTH	The data in a month
DAYOFWEEK	Day of the week
FLOOR	Round a number down
HOUR	Get the hour out of a datetime
LENGTH	Length of a string in bytes
LOWER	Return a string in lowercase
LOCATE	Return the index of the first occurrence of a substring in a larger string
LPAD	Pad a string to a given length by adding a particular character to its left
NOW	The current datetime
POW	Raise a number to a power
REGEXP	Whether a string matches a regular expression
REPLACE	Replace all occurrences of a particular substring in a string with a different substring
SQRT	Square root of a number
TRIM	Strip spaces from both sides of a string
UPPER	Return the upper-case version of a string

Besides just selecting rows/columns and operating on them, it is possible to aggregate many rows into a single returned value using a GROUP-BY statement. For example, this query will find the number of people named Helen in each state.

```
SELECT state, COUNT(name)
FROM people
GROUP BY state
WHERE first_name='Helen';
```

The COUNT() function used here is just one of many aggregator functions, which condense one columns from many rows into a single value. Several others are listed in the following table:

Function name	Description
MAX	Max value
MIN	Min value
AVG	Average value
STDDEV	Standard deviation

Function name	Description
VARIANCE	Variance
SUM	Sum

It is also possible to group by several fields, as in the following query:

```
SELECT state, city, COUNT(name)
FROM people
GROUP BY state, city
WHERE first_name='Helen';
```

A final word about basic queries is that you can give names to the columns you select, as in the following query:

```
SELECT state AS the_state,
 city AS where_they_live,
 COUNT(name) AS num_people
FROM people
GROUP BY state, city
WHERE first_name='Helen';
```

You could also have only renamed some of the columns. This renaming within the SELECT clause doesn't have any effect if all you're doing is pulling the data out. However, it becomes extremely useful if you are writing the results of your query into another table with its own column names or if you are working with several tables in the same query. More on those will follow later.

14.1.2 Joins

The final ingredient in the query language is the ability to join one table with another, which is a complicated enough topic that I'm giving it its own section. In a join, several tables are combined into one, with rows from the input tables being matched up based on some criteria (usually having specified fields in common, but you can use other criteria too). Joining is illustrated in this query, which tells us how many employees of each job title live in each state:

```
SELECT p.state, e.job_title, COUNT(p.name)
FROM people p JOIN employees e
ON p.name=e.name
GROUP BY p.state, e.job_title;
```

There are two things to notice about the new query. First, there is a JOIN clause, giving the table to be joined with people, and an ON clause, giving the criteria for when rows in the tables match. The second thing to notice is that "people p" and "employees e" give shorter aliases to the tables, and all columns are prefixed by the alias. This eliminates ambiguity, in case columns of the same name occur in both tables.

Every row under people will get paired up with every row under employees that it matches in the final table. So, if 5 rows under people have the name Helen and 10 rows under employees have the name Helen, there will be 50 rows for Helen in the joined table. This potential for blowing up the size of your data is one reason that joins can be very costly operations to perform.

The aforementioned query performs what is called an "inner join." This means that if a row under people does not match any rows under employees, then it will not appear in the joined table. Similarly, any row under employees that does not match a row under people will be dropped. You could instead have done a "left outer join." In that case, an orphaned row under people will still show up in the joined table, but it will have NULL in all the columns that come from the employees table. Similarly, a "right outer join" will make sure that every row under employees shows up at least once.

Outer joins are extremely common in situations where there is one "primary" table. Say, you are trying to predict whether a person will click on an ad, and you have one table that describes every ad in your database, what company/product it was for, who the ad was shown to, and whether it got clicked on. You might also have a table describing different companies, what industries they are in, and so on. Doing a left outer join between the ads and the companies will effectively just be adding additional columns to the ad table, giving new features that you might want to train a classifier on or calculate correlations between. Any companies that you didn't show ads for are superfluous for your study, and so get dropped, and any ad for which you happen to be missing the company data still stays put in your analysis. Its company-related fields will just be NULL.

14.1.3 Nesting Queries

The key operations in MySQL are SELECT, WHERE, GROUPBY, and JOIN. Most MySQL queries you will see in practice will use each of these at most once, but it is also possible to nest queries within each other.

This query takes a table of employees for a company, performs a query that counts how many employees are in each city, and then joins this result back to the original table to find out how many local coworkers each employee has:

```
SELECT ppl.name AS employee_name,
 counts.num_ppl_in_city-1 AS num_coworkers
FROM (
 SELECT
  city,
```

```
  COUNT(p.name) AS num_ppl_in_city
 FROM people
 GROUP BY p.city
) counts
JOIN people ppl
ON counts.city=ppl.city;
```

The things to notice in this query are as follows:

1) The subquery is enclosed in parentheses.
2) We give this result of the subquery the alias "counts," by putting the alias after the parentheses. Many SQL varieties require that an alias be given whenever you have subqueries.
3) We used the inner SELECT clause to give the name num_ppl_in_city to the generated column, as described earlier. This made the overall query much more readable.

14.1.4 Running MySQL and Managing the DB

The previous section described the syntax of MySQL queries. Now let's go over the nuts and bolts of how to access the servers, create/destroy tables, and put data in them.

There are many, many ways a MySQL server can be accessed. The easiest though is by the command line. The following command will open up an interactive session with a remote MySQL server:

```
mysql --host=10.0.0.4 \
  --user=myname \
  --password=mypass mydb
```

In this case, the user and password are your login credentials, mydb is the database on the server you want to access (although you can switch to a different later with a USE statement), and the host is the url of the server (since usually, the MySQL server is a physically different computer compared to the one you're working on).

Once you have opened up the MySQL shell, the following commands will let you do what you need to do. Their syntax should be straightforward. Note that each command is terminated by a semicolon. It is fine to break a single command out across several lines, so long as it ends with the semicolon.

```
# First list tables in current db
SHOW TABLES;
USE my_other_database; # switch to my_other_database
DROP TABLE table_to_drop; # delete this table
CREATE DATABASE my_new_database; # create a database
```

The syntax for creating a table is somewhat more complicated. Here is a simple example:

```
CREATE TABLE MyGuests (
id INT,
firstname VARCHAR(30),
lastname VARCHAR(30),
arrival_date TIMESTAMP
);
```

We say CREATE TABLE, the table's name, and then in parentheses we give the column names and their types. Note in this example that VARCHAR(30) means a string of at most 30 characters. This query will create an empty table, which we can then fill in a number of ways. If we only want to add a few entries, we can say

```
INSERT INTO MyGuests (id, firstname,
 lastname, arrival_date)
VALUES
(0, 'Field', 'Cady', '2013-08-13'),
(1, 'Ryna', 'Cady', '2013-08-13');
```

More often, we will want to load data from a CSV file or something of that nature, which can be done in the following way:

```
LOAD DATA INFILE '/tmp/test.txt'
INTO TABLE test
FIELDS DELIMITED BY ',';
```

When you are getting data out of a database by use of a query, it is also common to do it remotely from the command line. The following line of bash code will send a query to a MySQL server (which is assumed to be local host in this case, that is, the same computer you're sending the command from) and then write the results to a local CSV file:

```
mysql --host localhost \
-e 'SELECT * FROM foo.MyGuests' \
> foo.csv
```

14.2 Key-Value Stores

The next big database paradigm to know is the key–value store. This is conceptually similar to a Python dictionary on a massive scale, mapping keys (typically strings) to arbitrary data objects. This is fantastic for storing unstructured data such as web pages, and key–value stores have spiked in popularity in recent years, playing a particularly large role in the Big Data movement. The downside though is that key–value stores typically just give you access to the data; they don't have any of the optimized preprocessing that's available with an RDB.

In many cases, a key–value store is a more efficient way to store relational data in which most fields are blank. For example, let's say we have a database of people and know a few pieces of information about each one – many thousands of things that we could know about each person, but usually only a few of them are not NULL for any given person. In that case, it is more efficient to have the person/column tuple be a key and the entry be the value. If the database doesn't contain an entry for the person/column, you can assume that it's NULL.

Examples of key–value stores include Oracle NoSQL Database, redis, and dbm.

14.3 Wide Column Stores

A wide column store is a table such as a relational database, but it has a huge number of columns and is mostly sparse. Unlike relational databases, new columns can be added or deleted at will, and a given column will often exist for only a few related rows in the table. In this sense, a wide column store is perhaps more similar to a key–value store where the key contains two fields: a row id and a column name. Familiar MySQL operations, such as GROUPBY and JOIN, are often not supported.

The original wide column store is often considered to be Bigtable, an internal tool developed by Google and outlined in their paper *Bigtable: A Distributed Storage System for Structured Data*. Bigtable also has time stamps attached to each cell in the table and time-stamped records of all previous entries. The columns in a Bigtable are not completely independent of each other; they are grouped into "column families" that are frequently accessed together.

Open-source column-oriented databases include Cassandra, HBase, and Accumulo.

14.4 Document Stores

Document stores are similar to key–value databases, but the values they store are specifically documents of some flexible format, such as XML or JSON. In addition to simply fetching the data, document stores provide some functionality for searching through a database and pulling out documents (or parts of them) that match certain criteria or other processing. In this way, they can function as a happy medium between relational databases and key–value stores; they are flexible enough to hold nontabular data, but structured enough that they can actually process that data rather than just reading/writing it.

The most popular document store is MongoDB. It is an open-source, free piece of software produced by MongoDB Inc. To illustrate what document stores are like, I will give you a quick tutorial on MongoDB.

14.4.1 MongoDB®

Similarly to MySQL, data in Mongo is divided into databases. A database contains "Collections," the analogs of MySQL tables. A collection consists of a bunch of documents, stored in a JSON-like data format called BSON. A document in MongoDB is very flexible; they can mix and match fields, data types, layers of nesting, or whatever other variety you need to put in.

The only thing every document must contain is a unique identifier field called "_id," which functions as the primary key of the document. The user can specify the _id when a document is inserted; if there is another document with the same _id, you will get an error. Documents written into MongoDB without identifiers will be assigned them, with an object called an ObjectID. An automatically generated ObjectID will encode the time (in seconds) the ObjectID was created (typically, the time when its associated document was added to Mongo), a code for the computer it was generated on (very useful in cluster computing), and the process id that generated the ObjectID. Finally, each process that is generating ObjectIDs will keep track of how many it has created and encode the creation order in them, so that ObjectIDs created by the same process within one second of each other will still be distinct.

The main part to accessing data in MongoDB is the query. The following shell session will open a Mongo database, write some data to it, and execute a simple query to find all matching documents:

```
$ mongodb # command to open up a MongoDB shell
> // specify the database you want to use, like MySQL
> use some_db
> show collections
> // Insert a new document into the Collection "posts"
> // If posts doesn't exist, it will be created
> db.posts.insert({"name":"Bob","age":31})
WriteResult({ "nInserted" : 1 })
> // prints out the collections in this db.
> // Posts already exists.
> db.posts.find() // no query = show all documents
{ "_id" : ObjectId("55f8cc2f73b593e9ca69c126"), "name"
: "Field" }
```

```
> // shows one document matching given query
> db.posts.find({"name":"Bob"})
```

The query here looks like a JSON object itself. In a way it is: it maps the name of fields to all requirements we are placing on that field. In the aforementioned example, we wrote "name":"Bob", which means that "Bob" must be the name field in all matching documents. More generally, the query could have included a number of different constraints on the field, expressed in their own JSON-like form. Here are some additional examples:

```
> db.foobar.insert({"name":"Field","age":31})
> // age greater than 10
> db.foobar.find({"age":{$gt:10}})
> // age between 10 and 20
> db.foobar.find({"age":{$gt:10, $lt:20}})
> // Don't care what the age is, so long as it's there
> db.foobar.find({"age":{$exists:true}})
```

The expressions starting with "$" are called "query operators," and they constitute the logic of MongoDB queries. Before you call me on it, the query {"name":"Bob"} is just syntactic sugar around {"name":{$eq:"Bob"}}. A given MongoDB query will ultimately be a nested, JSON-like object, where the lowest level of nesting maps field names to query operators.

Besides the query, the find function takes an optional second argument called the projection. This specifies which fields on a document are to be returned (after all, the entire document might be quite large) and, if a document contains long arrays, how many elements of the arrays should be selected. The following table shows some example projections:

Projection	Meaning
{ _id:0, name:1, age:1}	Return only the name and age fields for each document. Note that we had to explicitly exclude _id; otherwise, it would be returned too
{ comments: { $slice: 5 } }	Return the first five elements of the "comments" field, which is presumed to be an array
{ comments: { $elemMatch: { userid: 102 } } }	Return all elements in the comments array that have userid 102

If we want to update documents in a collection, we use the update() command. It takes in two required arguments and a third optional one:

• A query statement indicating the documents to be modified. This uses the same query operators as the find() function.

- An update statement, with update operators that indicate what operations should be done to what fields.
- Optionally, a set of additional options. The most important additional option is "multi," which indicates whether multiple documents can be modified. It defaults to false, which is perhaps counterintuitive.

These commands give the basic idea:

```
> // This command updates a single document whose
"item" field is "ABC".
> db.inventory.update(
{ item: "ABC" },
{ $set: { "details.model": "14Q2" } })
> // This one will find all docs in the "clothing"
category, rename their category to "apparel",
> // and increment their "age" field by 5
> db.inventory.update(
{ category: "clothing" },
{$set: { category: "apparel" },
$inc: { age: 5 }},
{ multi: true })
```

14.5 Further Reading

1 Redmond, E, Wilson, J, *Seven Database in Seven Weeks: A Guide to Modern Databases and the NoSQL Movement*, 2012, Pragmatic Bookshelf, Raleigh, NC.
2 Tahaghoghi, S, Williams, H, *Learning MySQL*, 2006, O'Reilly Media, Newton, MA.

14.6 Glossary

BSON A JSON-like data format used by MongoDB.
Database A piece of software that stores data of a particular type in a format that supports low-latency access.
DB Common shorthand for database.
Document store A DB that stores documents, usually in a markup language such as XML or JSON.
Key–value store A DB that stores data objects by key but doesn't usually have other querying functionality.
MongoDB A popular open-source document store that runs on a cluster.
MySQL An extremely popular open-source version of SQL.

Query language A lightweight language for specifying database queries. Most query languages are based on SQL's syntax.

Relational algebra An idealized mathematical framework describing the operations of relational databases. Real-world RDBs typically include functionality that is not present in pure relational algebra.

Relational database A database that stores data in tables and supports operations such as selecting columns and joining records from several tables.

SQL The most popular relational database.

Wide column store Conceptually similar to a relational database, but typically tables have many, many columns, and they do not support operations such as joins and grouping.

15

Software Engineering Best Practices

In my experience, the single most important skill that is often lacking in data scientists is the ability to write decent code. I'm not talking about writing highly optimized numerical routines, designing fancy libraries or anything like that: just keeping a few hundred lines of code clear and manageable for the course of a project is a learned skill. I've seen many brilliant data scientists coming from areas such as physics and math, who lack this skill because they never had to write anything longer than a few dozen lines, or they never had to go back to code and update it. There's nothing worse than seeing a mathematical genius's data science project go up in flames because their 200-line script was so illegible they couldn't debug it.

That's one reason this chapter exists: to pass on the message that everybody who writes code is responsible for making sure that their code is clear.

The other reason for this chapter is that, in practice, data scientists are often called on to do far more than keep their code readable. Some companies have their data scientists focused strictly on analytics work. In many cases though, it falls to data scientists to turn their one-off scripts into reusable data analysis packages that take on a life of their own. Other times, data scientists function as junior members of a software engineering team, writing large pieces of production code that implement their ideas in a real-time product. This chapter will give you a condensed version of what goes into production-level code and what life is like on a software engineering team.

15.1 Coding Style

Coding style is not about being able to write code quickly or even about making sure that your code is correct (although in the long run, it enables both of these). Instead, it is about making your code easy to read and understand. This makes it easier for other people (including, most importantly, your future self

The Data Science Handbook, First Edition. Field Cady.

after you've forgotten how your code works) to figure out how your code works, modify it as need be, and debug it. Often, it takes a little longer to write your code well, but it is almost always worth the cost.

Let me give you a quick idea of the difference between good and bad code quality. Here is a piece of example code from earlier in the book which, if I may say so myself, I did a pretty good job of writing:

```python
from HTMLParser import HTMLParser
import urllib

TOPIC = "Dangiwa_Umar"
url = "https://en.wikipedia.org/wiki/%s" % TOPIC
class LinkCountingParser(HTMLParser):
    in_paragraph = False
    link_count = 0
    def handle_starttag(self, tag, attrs):
        if tag=='p': self.in_paragraph = True
        elif tag=='a' and self.in_paragraph:
            self.link_count += 1
    def handle_endtag(self, tag):
        if tag=='p': self.in_paragraph = False

html = urllib.urlopen(url).read()
parser = LinkCountingParser()
parser.feed(html)
print "there were", parser.link_count, \
    "links in the article"
```

There are no comments or documentation in this code, but it's fairly clear from my variable names what I'm trying to do, and the logical flow of the program shows how I'm doing it. Some of what I'm doing is specific to the HTMLParser library, but even if you aren't familiar with it, you can infer a lot about how it works from context. If you're familiar with HTML and the HTMLParser library, the code should be crystal clear.

In contrast, here is the same code, incorporating some of the no-nos of coding style:

```python
import urllib
url = "https://en.wikipedia.org/wiki/Dangiwa_Umar"
cont = urllib.urlopen(url).read()
from HTMLParser import HTMLParser
class Parser(HTMLParser):
    def handle_starttag(self, x, y):
        if x=='p': self.inp = True
        if x=='a':
```

```
        try:
          if self.inp:
            self.lc += 1
          except: pass
        if x=='html': self.lc = 0
    def handle_endtag(self, x):
        if x=='p': self.inp = False
p = Parser()
p.feed(cont)
print "there were",  p.lc, \
    "links in the article"
```

Among the sins I have committed, here are the following:

- Meaningless variable names.
- Hard-coding the url, rather than having the topic of the Wikipedia article called out as its own parameter.
- "Parser" is a very unhelpful class name.
- The Parser has two internal variables that it maintains: inp and lc (in_paragraph and link_count in the good code). These variables should have been declared upfront with default values, but instead I only create them on the fly as soon as I need them.
- The try–except loop is being used to check whether self.inp has been defined yet. That's a perverse use of try–except that will confuse people.
- In handle_starttag, there are three if-statements. But they are mutually exclusive, so it would have been clearer to use elif.
- The try–except loop is indented an irregular amount.

Good coding style isn't a set of rules that I can explain to you. It's really a mindset that you develop, an impatience with code being hard to read. The only way I know to learn it is by reading and writing a lot of code, so I won't try to teach it to you in this chapter. However, a few basic principles are as follows:

- Always use descriptive names for variables and functions.
- Modularity. Break the program down into self-contained parts that have clear functions.
- Avoid having excessive indentation, such as loops within loops within loops. If you're running into this, try to break the inner parts out into a separate function or subroutine.
- Use comments when appropriate. If you do something bizarre in your code, explain why in a comment. Also, every file should ideally start with a comment that explains what it does.
- Don't overuse comments. They distract from the code itself, which *should* be mostly clear enough that you can read it directly. Comments can also be wrong or out of date, so leave them out if the code speaks for itself.

- If you have several blocks of code that do very similar things, it often pays to refactor them into a single routine.
- Try to separate boilerplate information (such as the way data is formatted) from the core processing logic of the program. For example, the order of columns in a table should usually be specified separately from the code, which processes the table.

Some people might protest: if data scientists have to do all of this code quality stuff, then what's the difference between data science code and production code? My take is this: the goal of general code quality is to make it *easy* for somebody to read and understand your code. In contrast, the aim of production code is to make this understanding *unnecessary*. Production code is largely about creating APIs that are intuitive to use, well documented, and flexible enough to support a variety of use cases. People who use production code should be able to call it as a library without worrying about how it works. With data science programming, on the other hand, if somebody wants to modify your script, then it is fair to expect them to roll up their sleeves and dive into the code itself.

15.2 Version Control and Git for Data Scientists

A central feature of writing production software is a good version control system. This is a software framework that tracks changes that you make to a codebase, syncing them with a master copy stored on a server somewhere. It gives you the following massive advantages:

- If your computer gets destroyed, all of the code is backed up.
- If you're working on a team, everybody can keep their changes synced up by periodically reading down changes from the server.
- If things break, you can go back to a previous version of the codebase that was known to work.

The larger and more complex a codebase becomes, the more these become absolutely indispensable.

Typically, version control works by downloading the codebase as a directory on your computer. Editing the codebase is as simple as making changes to the contents of that directory (changing files, adding new files or subdirectories, etc.) and then telling the version control system to sync those changes. A version control system will afford at least the following functions:

- You can download the master copy of the codebase. This is often called a "checkout" or a "clone."
- You can refresh the code on your computer, incorporating any changes that have been made to the master copy since you originally check out the code. This is often called a "pull."

- After editing the code (changing it, adding new files, etc.), you can write your changes back to the master copy on the server. This is sometimes called a "commit," and sometimes a "push."
- If your changes conflict with changes that somebody else made, there is some way to "merge" the changes.
- You can include comments associated with your commits that describe what you did.
- Multiple users can check out the code and make changes, from multiple computers.
- It keeps track of the history of all changes.
- In case something gets broken, you can revert specific sets of changes.

Most modern systems also include the ability to create branches of the entire codebase. If you have a large amount of work that needs to be done, it is often useful to fork off a new branch of the entire codebase. This lets you do whatever you need to do without worrying about polluting the master branch. Later, when all changes have been made, you can merge your changes back into the master branch and deal with any possible conflicts.

Branches are also useful if you need to make one-off versions of the codebase for some special purpose. For example, maybe there is a customer who wants a customized version of your product. This may involve small changes to many different parts of your codebase and require its own version of the branch.

The most popular version control system these days is Git, although there are others. The following table is a cheat sheet of the most important git commands, if you use git form in the command line. All of them except the first are expected to be executed when you are inside the checked-out directories.

Sample command	What it does
git clone https://github.com/someproject.git	Clones the master branch of the repo
git status	Says what files you have modified or added
git add myfile.py	Makes a change by adding a new file
git rm myfile.py	Makes a change by deleting a file
git mv dir1/file.py dir2/file.py	Makes a change by moving a file
git commit myfile.py –m "my commit message"	Any changes you have made to the file (including adding it or deleting it) are staged for pushing
git push	Pushes all committed changes to the master copy of the branch you are on
git diff myfile.txt	Displays all changes that I've made to the file that have not been committed yet

Sample command	What it does
git branch	Lists all branches of the repo that your local git knows about and can switch to
git checkout my_branch	Switches to the branch called my_branch
git branch my_new_branch	Creates a new branch called my_new_branch and check it out
git merge other_branch	Merges the changes made in other_branch into the branch you're currently on

The other notable thing about git is the website www.github.com. Github allows anybody to create a Git repo where the master copy is stored on Github's internal servers. Github is free to use provided that you make the codebase open to the public, so there are many fascinating projects hosted on it. Many people (yours truly included) will also use github repositories as a place to store some of their personal code snippets. For a fee, Github will keep a repository private, so it's a great way for small companies to get their hands on world-class version control.

15.3   Testing Code

There's a spectrum of how rigorously code can be tested. Scientists and mathematicians are typically used to "testing" code in a pretty casual way – making sure that it seems to give the correct output, as indicated by a few sanity checks. On the other end of the spectrum, large software projects depend on complicated testing frameworks that can sometimes be as complicated as the source code itself. Data science tends toward the former, but in practice, it can run the gamut. This section will review some of the standard testing concepts that you would see in a hardcore software environment.

At first glance, writing and maintaining testing code might appear to be an irritating burden when you have source code to write. It's a pain to go back and revise all the tests to reflect the changes you have made, rather than going on to the *other* changes you have to make. It's even possible that all of your main code works perfectly, but the tests fail because there was a bug in your testing code. Who wants to deal with that?

These are understandable concerns, and they are sometimes valid for small code bases. However, as the number of lines increases, and especially as the software is maintained for a long term rather than being one-off scripts, a robust testing framework becomes absolutely vital. It took me a long time to appreciate its value, but believe me; it's there.

The most obvious advantage of testing code is, of course, that it checks whether your code is right. There is another, less obvious bonus too. Oftentimes, the test code is the best documentation of your source code. Rather than describing your code's functionality in nebulous English, it shows the brass tacks of how to call your code, which inputs go in, and which outputs are generated. Best of all, by simply running all the tests, you can make sure that this "documentation" is up to date.

15.3.1 Unit Tests

Unit tests cover small, self-contained pieces of code logic. For a given piece of code, its unit tests should cover every known edge case of the code, as well as several of the more general cases. Many programming languages have libraries that are specifically designed to support unit testing.

Let's see how this looks with the Python unit testing library. The following code assumes that you've written a library called "mymath" containing a function "fib", which calculates the nth Fibonacci number. This code assumes that the zeroth Fibonacci number is 0, the first is 1, and every subsequent number is the sum of the previous two:

```python
import unittest
from mymath import fib

class TestFibonacci(unittest.TestCase):
  def test0(self):
    self.assertEqual(0, fib(0))
  def test1(self):
    self.assertEqual(1, fib(1))
  def test2(self):
    self.assertEqual(fib(0)+fib(1), fib(2))
  def test10(self):
    self.assertEqual(fib(8)+fib(9), fib(10))

unittest.main()
```

The TestCase class is where all of the magic happens. For the given piece of code you want to test, you create a class that inherits from it, in this case, TestFibonacci. For every test you want to run on the code, you create a new method, whose name must start with "test." These tests use the assertEqual function, which is a member function of Test Case and keeps track of any failures. The main() function looks for all test cases defined in the namespace, runs them all, and reports the results to the user.

TestCase has a number of other methods, some of which are listed in this table:

assertNotEqual	Make sure that two things are different
assertTrue	Make sure that a variable is True
assertFalse	Make sure that a variable is False
assertRaises	Make sure that calling some function raises an error

The niftiest application of unit tests is in long-term code maintenance. Say that you wrote a code module a while ago, and a small glitch gets discovered. Fixing the glitch might involve substantial changes to your source code, and you run the risk of breaking the functionality that was already there. So, you keep all your old unit tests in place and add one that tests the new edge case. As soon as the new test passes, and all of the previous tests *still* pass, you can be confident that you've fixed the problem without breaking anything.

Another great use for unit tests comes up if you are using git and working with other people. Say you've made some changes to the codebase and want to make sure that everything still works when combined with changes other people have made. You can use the "git pull" command to pull down all their changes, recompile the code if need be, and rerun the unit tests. If they all pass, you can push your changes with confidence.

Personally, I like to look at source code and test code as two symbiotic halves. They are "in sync" if all of the unit tests pass. The source code is dangerously likely to have a bug, and the test code is dangerously likely to have a bug. But a bug in either will break the symbiosis. So, if the source code and test code are in sync, your code almost certainly does what it's supposed to do. The probability that both halves have a bug, and those bugs cancel out so that the tests pass, is miniscule. All you have to do is make sure that all of the edge cases in the source code are properly tested.

The key limitation of unit tests is that each code module is tested in isolation. Unit testing isn't really intended for code that accesses external resources, such as the Internet, a remote MySQL server, or other processes running on your computer. If you must unit test that kind of software, the standard way to do it is with something called a "mock." If your code is designed to access some external API, a mock is an object that mimics that API in a predefined way. For example, a mock of a MySQL server might take in queries but always returns a predefined result, without ever trying to access a remote server.

Now of course, you *can* write a Python script that uses the unittest library but also accesses external resources. This is bad practice: the tests might fail because the external resource is having trouble, or it isn't available from the computer that is running the tests. But if you want to run unit tests manually as a way to test your own code, calling the external resources explicitly is often easier than creating mocks.

15.3.2 Integration Tests

In integration tests, the different code modules are linked up to each other, external resources are put in place, and the system is run in a realistic way. This is where you run into network timeouts, permissions glitches, memory overflows, and other errors that can only occur at scale. It's also the trial-by-fire of whether every module correctly understands every other module's API.

There is usually not a standard library for integration testing, because it is so specific to each individual project. It's more of a stage in the development of a serious piece of software.

15.4 Test-Driven Development

Previously, I explained how you can use unit tests to make sure that you've fixed a glitch in your code without breaking something that was already working. "Test-driven development" (TDD) is an approach to software engineering that takes this idea to the extreme. You write up the unit tests for your module before you even start on the source code. Then, your goal in writing the module is just to make the unit tests all pass. Often you address each test in turn, making sure that it passes while not breaking the others, but ignoring any test that you haven't come to yet.

There are two big advantages to TDD. The first is that it makes you think through your module's desired functionality right upfront. This forces you to decide on a preliminary API and makes you write code that calls the API so that you can see if it's painfully clunky. The second advantage of TDD is that, when it comes to writing the source code itself, you are able to laser-focus on a single test rather than trying to hold the entire system in your head at once. It can be very calming, almost meditative.

TDD is also only appropriate if you know more or less what your software should ultimately do. This is often not the case in data science, because you don't know what feature extractions and preprocessing will work until you start playing with the data. In my own experience, data science is often used to figure out exactly which analyses should be performed in what way, and then TDD is used to implement a production version.

The other issue with TDD is that sometimes its myopia is not practical. The more different modules interact with each other, the more important it becomes to plan out your software architecture ahead of time. Otherwise, each additional unit test may require rejiggering pretty much your entire code base, and you could run into showstopper issues when you think you're most of the way through the project.

15.5 AGILE Methodology

Test-driven development is a way for individual programmers to go about getting their work done. Agile, on the other hand, is a way to organize teams of developers. The term was coined in 2001 in "The Manifesto for Agile Software Development." The book was written by a group of programmers who were sick of top-down, long-term plans resulting in projects that ultimately fail. The concept caught on like wildfire, and some variant of agile is ubiquitous in data science and software.

The key idea of AGILE is to make projects more flexible by shortening the development cycle and tightening feedback loops. Among the key principles of agile are the following:

- Frequent collaboration and decision-making by the individual team members, as opposed to mandates from above
- Similarly frequent communication with clients and stakeholders
- Making sure that there is always a working end-to-end product, even if it doesn't include all of the features that you will ultimately want.

A typical feature of agile teams is a daily morning meeting, often called "stand-up" or "scrum." Typically, this will consist of the team going around in a circle, with every member saying (1) what they accomplished the previous day, (2) what they are planning to do today, and (3) any roadblocks they are running into.

Agile is often a fantastic way to approach software, but it is not without its own drawbacks. The biggest is that it sometimes comes at the expense of long-term planning and clear direction. The second issue with agile development is that the focus on rapid feature iteration often leads to an accumulation of "technical debt" – disorganization and instabilities in the codebase that come back to bite you later.

15.6 Further Reading

1 Rubin, K, *Essential Scrum: A Practical Guide to the Most Popular Agile Process*, 2012, Addison-Wesley, Boston, MA.
2 Martin, R, *Clean Code: A Handbook of Agile Software Craftsmanship*, Prentice Hall, Upper Saddle River, NJ.

15.7 Glossary

AGILE development An approach to software development that focuses on getting a minimal product working quickly and then having short "sprints" with clear goals that are incremental improvements.
Git The most popular version control system today.
Integration test A test that makes sure that several pieces of software work correctly together.
Perl Golf A pejorative term for cramming a lot of functionality into a few lines of code, making it hard to understand.
Scrum A short, daily team meeting in AGILE development.

Sprint A short period of time (usually 2 weeks) over which concrete goals are set in AGILE development.

Technical debt Work that will need to be done in the future because of shortsighted and/or expedient programming choices made in the past.

Test-driven development A system of writing code where you start by writing the unit tests for whatever changes you intend to make and then changing the code until the new unit tests pass and all the old ones still pass.

Unit test A test that runs a small piece of code in isolation and makes sure that it works correctly. They are particularly useful for test-driven development and for testing edge cases in the code's logic.

Version control A piece of software that tracks changes made by multiple users to a repository of code.

16

Natural Language Processing

Natural language processing (NLP) is a collection of techniques for working with human language. Examples would include flagging e-mails as spam, using Twitter to assess public sentiment, and finding which text documents are about similar topics. NLP is an area that many data scientists never actually need to touch. But enough of them end up needing it, and it is sufficiently different from other subjects that it deserves a chapter in this book.

This chapter will start with several generic sections about NLP datasets and big-picture concepts. Then I will switch gears to the core NLP concepts, moving from the simple, quick-and-dirty techniques to more complicated ones.

I also want to emphasize that NLP techniques are not strictly limited to language. I've also seen them used to parse computer log files, figuring out what "sentences" the computer generates. Personally, I first learned many of the statistical techniques while working with bioinformatics.

16.1   Do I Even Need NLP?

The first question to ask when using NLP is whether you even need it. There is often pressure from customers and bosses to solve problems using NLP, because it is seen as some kind of magical silver bullet. But in my experience, NLP is hard to implement, and it is prone to bizarre errors that are obviously wrong when a human looks at them.

I've seen people bang their heads against a problem using NLP techniques, only to eventually give up and try solving the problem with regular expressions. Then lo and behold, the regular expressions work better than the NLP ever did.

Here are a few thoughts to keep in mind when deciding whether to try NLP:

- If your data has a regular structure to it, then you can probably extract what you need without NLP.

The Data Science Handbook, First Edition. Field Cady.
© 2017 John Wiley & Sons, Inc. Published 2017 by John Wiley & Sons, Inc.

- NLP tends to be very effective at tasks such as determining whether two documents have similar content, because simple things such as word frequency are very informative.
- If you are trying to extract facts from documents, it is almost impossible unless you have standardized language, such as Wikipedia or legal contracts. You probably can't do it with something such as Twitter.
- It is often hard to make sense of why an NLP algorithm performed in the way it did.
- NLP typically requires a lot of training data.

16.2 The Great Divide: Language versus Statistics

There are two very different schools of thought in NLP, which use very different techniques and sometimes are even at odds with one another. I'll call them "statistical NLP" and "linguistic NLP." The linguistic school focuses on understanding language as language, with techniques such as identifying which words are verbs or parsing the structure of a sentence. This sounds great in theory, but it is often staggeringly difficult in practice because of the myriad ways that humans abuse their languages and break the rules. The statistical school of NLP solves this problem by using massive corpuses of training data to find statistical patterns in language. They might notice that "dog" and "bark" tend to occur frequently together or that the phrase "Nigerian prince" is more common in a corpus of e-mails than chance would dictate. Personally, I see statistical NLP mostly as a blunt-force workaround for the fact that linguistic NLP is so extraordinarily difficult.

In the modern era of massive datasets (such as the web), this divide has become more pronounced, and statistical NLP tends to have the advantage. The best machine translation engines, such as the ones Google might use to automatically translate a website, are primarily statistical. They are built by training on thousands of examples of human-done translation, such as newspaper articles published in multiple languages or books that were translated. Some linguists protest that this is dodging the scientific problem of figuring out how the human brain really processes language. Of course it *is*, but the bottom line is that the results are generally better. On the other hand, training something like that requires a training corpus and specially crafted machine learning algorithm that is generally beyond the reach of all but the most sophisticated NLP users.

16.3 Example: Sentiment Analysis on Stock Market Articles

The following script shows off a tiny taste of what is possible with Python's most popular NLP library: nltk ("natural language toolkit"). You set the ticker symbol of a stock near the top of the script. It then parses a bunch of recent articles about the stock, gauges them as positive or negative, and prints how many fell into each category.

```
import re
import urllib
import nltk
from nltk.corpus import wordnet as wn
from nltk.corpus import stopwords
from nltk.stem.wordnet import WordNetLemmatizer
TICKER = 'CSCO'
URL_TEMPLATE = "https://feeds.finance.yahoo.com/" + \
    "rss/2.0/headline?s=%s&region=US&lang=en-US"
def get_article_urls(ticker):
    # Return list of URLs for articles about a stock
    link_pattern = re.compile(r"<link>[^<]*</link>")
    xml_url = URL_TEMPLATE % ticker
    xml_data = urllib.urlopen(xml_url).read()
    link_hits = re.findall(link_pattern, xml_data)
    return [h[6:-7] for h in link_hits]
def get_article_content(url):
    # input: url for a news article
    # output: approx. content of the article
    # Downloads HTML for an article and then
    # pulls data from paragraphs in the html
    paragraph_re = re.compile(r"<p>.*</p>")
    tag_re = re.compile(r"<[^>]*>")
    raw_html = urllib.urlopen(url).read()
    paragraphs = re.findall(paragraph_re, raw_html)
    all_text = " ".join(paragraphs)
    content = re.sub(tag_re, "", all_text)
    return content
def text_to_bag(txt):
    # Input: bunch of text
    # Output: bag-of-words of lemmas
    # Also removes stop words
    lemmatizer  = WordNetLemmatizer()
    txt_as_ascii = txt.decode(
        'ascii', 'ignore').lower()
    tokens = nltk.tokenize.word_tokenize(txt_as_ascii)
    words = [t for t in tokens if t.isalpha()]
    lemmas = [lemmatizer.lemmatize(w) for w in words]
    stop = set(stopwords.words('english'))
```

```
        nostops = [l for l in lemmas if l not in stop]
        return nltk.FreqDist(nostops)
def count_good_bad(bag):
    # Input: bag-of-words of lemmas
    # Output: number of words that are good, bad
    good_synsets = set(wn.synsets('good') + \
        wn.synsets('up'))
    bad_synsets = set(wn.synsets('bad') + \
        wn.synsets('down'))
    n_good, n_bad = 0, 0
    for lemma, ct in bag.items():
        ss = wn.synsets(lemma)
        if good_synsets.intersection(ss): n_good += 1
        if bad_synsets.intersection(ss): n_bad += 1
    return n_good, n_bad

urls = get_article_urls(TICKER)
contents = [get_article_content(u) for u in urls]
bags = [text_to_bag(txt) for txt in contents]
counts = [count_good_bad(txt) for txt in bags]
n_good_articles = len([_ for g, b in counts if g > b])
n_bad_articles = len([_ for g, b in counts if g < b])
print "There are %i good articles and %i bad ones" %
    (n_good_articles, n_bad_articles)
```

16.4   Software and Datasets

NLP processing is generally very computationally inefficient. Even something as simple as determining whether a word is a noun requires consulting a lookup table containing a language's entire lexicon. More complex tasks such as parsing the meaning of a sentence require figuring out a sentence's structure, which becomes exponentially more difficult if there are ambiguities in the sentence (which there usually are). This is all ignoring things such as typos, slang, and breaking grammatical rules. You can partly work around this by training stupider models on very large datasets, but this will just balloon your data size problems.

There are a number of standardized linguistic datasets available in the public domain. Depending on the dataset, they catalog everything from the definitions of words, to which words are synonyms of each other, to grammatical rules. Most NLP libraries for any programming language will leverage at least one of these datasets.

One lexical database that deserves special mention is WordNet. WordNet covers the English language, and its central concept is the "synset." A synset is a collection of words with roughly equivalent meanings. Casting every word to its associated synset is a great way to compare whether, for example, two sentences are discussing the same material using different terms. More importantly, an ambiguous word such as "run," which has many different possible meanings, is a member of many different synsets; using the correct synset for it is a way to eliminate ambiguity in the sentence. Personally, I think of synsets as the words of a separate language, one in which there is no ambiguity and no extraneous synonyms.

16.5   Tokenization

The first part of any NLP process is simply breaking a piece of text into its constituent parts (usually words). This process is called "tokenization," and it is complicated by issues such as punctuation markers, contractions, and a host of other things. Is "that's" one word, or is it two? If two, should the second word be "is" or "'s."

The process can become even more complicated if your tokens are sentences rather than words.

16.6   Central Concept: Bag-of-Words

Probably, the most basic concept in NLP (aside from some very high-level applications of it) is that of a "bag-of-words," also called a frequency distribution. It's a way to turn a piece of free text (a tweet, a Word document, or whatever else) into a numerical vector that you can plug into a machine learning algorithm. The idea is quite simple – there is a dimension in the vector for every word in the language, and a document's score in the nth dimension is the number of times the nth word occurs in the document. The piece of text then becomes a vector in a very high-dimensional space.

Most of this chapter will be about extensions of the bag-of-words model. I will briefly discuss some more advanced topics, but the (perhaps surprising) reality is that data scientists rarely do anything that can't fit into the bag-of-words paradigm. When you go beyond bag-of-words, NLP quickly becomes a staggeringly complicated task that is usually best left to specialists.

My first-ever exposure to NLP was as an intern at Google, where they explained to me that this was how part of the search algorithm worked. You condense every website into a bag-of-words and normalize all the vectors. Then when a search query comes in, you turn it into a normalized vector too, and then take its dot product with all of the web page vectors. This is called the "cosine similarity," because the dot product of two normalized vectors is just the cosine of the angle between them. The web pages that had high cosine similarity were those whose content mostly resembled the query, that is, they were the best search result candidates.

The majority of this chapter will be about extensions and refinements of the basic idea of bag-of-words – we will discuss some of the more intricate (and error-prone) sentence parsing toward the end. Right off the cuff, we might want to consider the following extensions to the word vector:

- There's a staggering number of words in English and an infinite number of potential strings that could appear in text. We need some way to cap them off.
- Some words are much more informative than others – we want to weight them by importance.
- Some words don't usually matter at all. Things such as "I" and "is" are often called "stop words," and we may want to just throw them out at the beginning.
- The same word can come in many forms. We may want to turn every word into a standardized version of itself, so that "ran," "runs," and "running" all become the same thing. This is called "lemmatization."
- Sometimes, several words have the same or similar meanings. In this case, we don't want a vector of words so much as a vector of meanings. A "synset" is a group of words that are synonyms of each other, so we can use synsets rather than just words.
- Sometimes, we care more about phrases than individual words. A set of n words in order is called an "n-gram," and we can use n-grams in place of words.

NLP is a deep, highly specialized area. If you want to work on a cutting-edge NLP project that tries to develop real understanding of human language, the knowledge you will need goes well outside the scope of this chapter. However, simple NLP is a standard tool in the data science toolkit, and unless you end up specializing in NLP, this chapter should give you what you need.

When it comes to running code that uses bag-of-words, there is an important thing to note. Mathematically, you can think of word vectors as normal vectors: an ordered list of numbers, with different indices corresponding to different words in the language. The word "emerald" might correspond to index 25, the word "integrity" to index 1047, and so on. But generally, these vectors will be stored as a map (in the aforementioned Python code, it is a FreqDist object) from word names to the numbers associated with those words. There is often no need to actually specify which words correspond to which vector indices, and doing so would add a human-indecipherable layer in your data processing, which is usually a bad idea. In fact, for many applications, it is not even necessary to explicitly enumerate the set of all words being captured. This is not just about human readability: the vectors being stored are often quite sparse, so it is more computationally efficient to only store the nonzero entries.

You might wonder why we would bother to think of a map from strings to floats as a mathematical vector. If so, fair question. The reason is that vector operations, such as dot products, end up playing a central role in NLP. NLP pipelines will often include stages where they convert from map representations of word vectors to more conventional sparse vectors and matrices, for more complicated linear algebra operations such as matrix decompositions.

16.7 Word Weighting: TF-IDF

The first correction to bag-of-words is the idea that some words are more important than others. In some cases, we know a priori which words to pay attention to, but more often we are faced with a corpus of documents and have to determine from it which words are worth counting in the word vector and what weight we should assign to them.

The most common way to do this is "Term Frequency–Inverse Document Frequency" (TF-IDF). The intuition behind TF-IDF is that rarer words are more important. You calculate the word vector for a particular document by counting up all of its words' frequencies, as usual. But then you divide each count by the frequency of that word in the training corpus. This will dampen down the scores for common words, but balloon them for rare words that happen to occur (comparatively) frequently in this document.

Generally in a TF-IDF, you will only look at words that have some minimal frequency in the training corpus. If a word never occurs in the training corpus but shows up in a new document, then surely it shouldn't have infinite importance. And we don't want to give aberrantly high importance to a word that shows up only once. Five occurrences in the corpus is a common minimum for the word to be counted. Even requiring just two will chop out a lot of noise such as typos.

16.8 *n*-Grams

Often we don't want to look just at individual words, but phrases. The key term here is an "*n*-gram" – a sequence of *n* words that appear consecutively. A piece of text containing M words can be broken into a collection of $M - n + 1$ *n*-grams, as shown in the following figure for 2-grams:

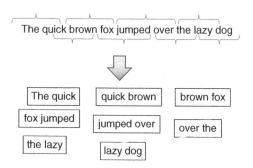

You can create a bag-of-words out of n-grams, run TF-IDF on them, or model them with a Markov chain, just as if they were normal words.

The problem with *n*-grams is that there are so many potential ones out there. Most *n*-grams that appear in a piece of text will occur only once, with the frequency decreasing, the larger *n* is. The general approach here is to only look at *n*-grams that occur more than a certain number of times in the corpus.

16.9 Stop Words

Bag-of-words, TF-IDF, and *n*-grams are fairly general processing techniques, which can be applied to many other areas. Now let's move into some more truly more NLP-oriented extensions to bag-of-words. In most cases, this consists of a preprocessing step that canonicalizes the text – putting everything in a standardized format suitable for downstream processing by something such as TF-IDF.

The simplest version is to remove what are called "stop words." These are words such as "the," "it," and "and" that aren't really informative in and of themselves. They are critically important if you're trying to parse the structure of a sentence, but for bag-of-words, they are just noise. There is no absolute definition of "stop words." Frequently, they are found by taking the most common words in a corpus, then going through them by hand to determine which ones aren't really meaningful. In other cases (such as the aforementioned example script), there is a list of them prebuilt into an NLP library.

Stop words become problematic when you are using *n*-grams. For example, the very informative phrase "to be or not to be" is likely to get stopped out entirely!

16.10 Lemmatization and Stemming

The other big approach is called "lemmatization." The "lemma" for a word is the base word from which it is derived. Intuitively, if we are making a bag-of-words, then "running," "ran," and "runs" should all count the same. In linguistic terms, they are all variations of the lemma "run," and we would want to turn them all into "run" as a preprocessing step.

Some form of lemmatization is extremely important in English, but it is critical in many other languages. I was surprised when I first learned that English is actually fairly tame when it comes to modifying our words. Languages such as Spanish, for example, have an arbitrary "gender" assigned to their nouns, and any adjective will be tweaked to reflect that gender of the noun it is describing. Other languages make liberal use of "case" in their nouns, where the noun is modified to reflect its part of speech. Noun case is the difference between "I," "me," "my," and "mine"; English uses noun case in a few special instances as this, but otherwise, the only common casing is the use of "'s" to indicate possession.

The problem with lemmatization is that, in general, it is extremely computationally expensive because it requires a certain amount of understanding of the text. Extracting the lemma for a given word requires knowing its part of speech, for example, which requires analysis of the surrounding sentence.

A simpler approach, which is less accurate than lemmatization but quicker to run and easier to implement, is called "stemming." The "stem" of a word is very similar to the lemma. The difference is that the lemma is itself a version of the word in question, whereas the stem is just the part of a word that doesn't change despite the inflection of the word. So between the words "produce," "producer," "product," "production," and "producing," the lemma will be "produc-".

The implementation of a stemmer generally just applies many different rules of thumb to try reducing things down to their stems. For example, it will typically strip off "ing" or "ation" if they occur at the end of a word (unless such stripping would leave too few remaining letters). I have even seen them implemented using a big lookup table, which maps the words of the language in their various forms to their respective stems.

16.11 Synonyms

Intuitively, when we are trying to analyze the text, the words themselves are less important than the "meaning." This suggests that when we might want to collapse related terms such as "big" and "large" into a single identifier. These identifiers are often called "synsets," for sets of synonyms. Many NLP packages leverage use of synsets as a major component of their ability to understand a piece of text.

The simplest use of synsets is to take a piece of text and replace every word with its corresponding synset. This is a souped-up version of lemmatization, because we don't just collapse "run" and "running" into a single thing – we also mix "sprinting" in there. Ultimately, I think of synsets as constituting the vocabulary of a "clean" language, one in which there is a one-to-one matching between meanings and words. It strips out all the ambiguity that confounds computer programs studying real languages.

That's all very rosy-sounding, but there is a major hitch: in general, a word can belong to several synsets, because a word can have several distinct meanings. So doing a translation from the original language to "synset-ese" is not always feasible.

16.12 Part of Speech Tagging

As we move from purely computational techniques toward ones based more closely on language, the next stage is part-of-speech tagging (POS tagging), where we identify whether words in a sentence are nouns, verbs, adjectives, and so on. This can be done in nltk as follows:

```
>>> nltk.pos_tag(["I", "drink", "milk"])
[('I', 'PRP'), ('drink', 'VBP'), ('milk', 'NN')]
```

In this case, the PRP tag tells us that "I" is a prepositional phrase, "drink" is a verb phrase, and "milk" is a common noun in singular form. A complete list of the POS tags nltk uses can be seen by calling

```
>>> nltk.help.upenn_tagset()
```

16.13   Common Problems

This section will give a brief overview of a number of the areas to which NLP can be applied. Each section is a massive subject in its own right, with a sophisticated suite of techniques and best practices. As a data scientist, you are unlikely to build a state-of-the-art version of any of them. However, you could easily be called upon to do a simple version, and if so, this section should give you some pointers.

16.13.1   Search

One of the most straightforward tasks for NLP is searching through a set of documents to find those that match a query.

Searching is often broken into two types: navigational and research. In navigational search, there is typically a single document that the user is trying to locate, and the goal of the search engine is to find it. Research search is much more general – typically, a user does not know which documents, if any, are relevant to their query, and they expect to be inspecting a lot of them by hand.

Anybody who has used a modern search engine can imagine the wealth of special cases, caching of common queries, and other work involved in creating such a product. It is well beyond the scope of what most data scientists do.

Searching is often implemented by extracting a bag-of-words for the query string and for every document in the corpus. Especially if you are baking complicated linguistic processing into your bag-of-words, vectorizing an entire corpus can be an extremely computationally intensive process. Fortunately though, you only have to do it once, and you can store the vectors for later use. Executing a search then just consists of vectorizing the query itself (which is typically much easier, because the query is short) and comparing its vector against all those in the database.

The typical way that vectors are compared is called "cosine similarity," which consists of the following steps:

1) Normalize each vector, so that the sum of squares of its numbers is 1.0. This can be done offline for your corpus of documents, so that you only have to normalize the query itself at query time.
2) Take the dot product of the query and each normalized corpus vector.

The resulting number is called the "cosine similarity" because, in the case of two vectors of length 1 (remember we normalized them), the dot product is just the cosine of the angle between them. It will be 1.0 if the vectors are identical, falling off to 0.0 if the query and the text have no words in common. Cosines can of course be negative, but in the case of bag-of-words, all components are nonnegative (since you can't have fewer than 0 occurrences of a word in a piece of text), so the lowest possible cosine similarity is 0.

16.13.2 Sentiment Analysis

Sentiment analysis is typically used to refer to gauging the tone of a piece of text – positive, negative, or neutral. This is what we did in the example script at the beginning of this chapter to identify, in a fraction of a second, whether analysts are saying good or bad things about the stock. Ideally, we can get this insight before any flesh-and-blood humans have a chance to read the article the old-fashioned way and trade based on it. There are more complicated versions of sentiment analysis that can, for example, determine complicated emotional content such as anger, fear, and elation. But the most common examples focus on "polarity," where on the positive–negative continuum a sentiment falls.

Simple sentiment analysis is often done with handmade lists of keywords. If words such as "bad" and "horrendous" occur a lot in a piece of text, it strongly suggests that the overall tone is negative. That's what we did in our example script.

Slightly more sophisticated versions are based on plugging bag-of-words into machine learning pipelines. Commonly, you will classify the polarity of some pieces of text by hand and then use them as training data to train a sentiment classifier. This has the massive advantage that it will implicitly identify key words you might not have thought of and will figure out how much each word should be weighted. If you use extensions of the bag-of-words model, similar to n-grams, you can also identify phrases such as "nose dive," which deserve a very large weight in sentiment analysis but whose constituent words don't mean much.

The most advanced sentiment analysis goes beyond bag-of-words. In this case, you must do things such as parsing a sentence to figure out which entities are being described with words such as "bad." However, you can often work around this by examining smaller pieces of text. If you are parsing an article on an industry in the stock market, for example, there are likely to be many companies discussed in both positive and negative ways. However, a given sentence or paragraph is likely to have one predominant sentiment and only refer to a single company.

16.13.3 Entity Recognition and Topic Modeling

In many cases, we have a corpus of documents and want to determine what "things" they talk about. This can run the gamut from pulling out specific entities (such as the names of human beings mentioned in a document) to more general topics.

Typically, identifying specific entities being discussed is referred to as "entity recognition" or "named entity recognition." There are many ways to go about this, and you can imagine the amount of hand coding (or massive amounts of training data) that is often required to recognize that "Robert" and "Bob" are likely to be the same person. Entity recognition often makes extensive use of POS tagging, because generally, it is only the nouns in a sentence (and usually, only the proper nouns at that) that are viable candidates for entities.

"Topic modeling" usually refers to finding much broader topics. A given piece of text is generally thought of as a combination of several topics, and the critical intuition is that words that are used frequently together tend to be related to the same "topic." For example, a document that is half about cats and half about dogs should have similar amounts of dog-related terms (bark, wolf, howl, etc.) and cat-related terms (purr, litter, etc.).

You might think that this sounds like an obvious place to apply principal component analysis (PCA). If so, then you're pretty close to the mark, but not quite hitting it. The usual tool of choice for topic modeling is called latent semantic analysis (LSA), and it is based on a mathematical notion called singular value decomposition (SVD). As with PCA, every "topic" in our corpus corresponds to a vector in word space, and we express every document as a linear combination of these topics. For instance, a particular topic might have large weight on words such as "touchdown," "quarterback," and "ball." The reason we use SVD rather than PCA is that SVD forces our components to be orthogonal to each other. This avoids a potentially awkward situation where one of our topic vectors can be expressed, as least partly, as a combination of other topics.

16.14 Advanced NLP: Syntax Trees, Knowledge, and Understanding

I promised that I would briefly discuss topics that go beyond what can be done with bag-of-words and require something in the direction of "understanding" the text. If, for example, you are trying to answer specific question about the data such as "who is John dating?," then you will need something beyond bag-of-words. Typically in these situations, NLP is used to create "knowledge bases," which store facts in a format where machines can use them for queries and reasoning.

The idea of a knowledge base isn't really new, and in fact, the mathematical theory of relational databases functioned essentially as a knowledge base that supported logical queries.

Typically, a knowledge base will have tables that represent facts. For example, you might have a table called IsParentOf, which has one column for a parent and another for their child. In this way, a knowledge base is very similar to a relational database, and logical questions become equivalent to SQL-like queries. For example, we could find all people who have at least two children by saying

```
SELECT a.parent
FROM IsParentOf as a
JOIN IsParentOf as b
ON a.parent = b.parent
WHERE a.child != b.child
```

You could use a relational database as a poor man's knowledge system. Combing through bodies of text you could, for example, identify every place that somebody is said to be the mother or father of somebody else and use this to populate the IsParentOf table. The problem with this is that we have to know going into it that fatherhood is a relationship we are interested in and how to parse it out. Additionally, there are other logical rules that are not captured in an RDB, such as that fact that while a parent can have many children, every child can have at the most two parents.

Modern knowledge bases typically augment the SQL-like tables with logical rules that describe the categories of things being discussed and the relationships between them. These collections of domain-specific categories and rules are also often called "ontologies."

16.15 Further Reading

1 Bird, S, Klein, E & Loper, E, *Natural Language Processing with Python*, 2009, O'Reilly Media, Newton, MA.

2 Jurafsky, D & Martin, J, *Speech and Language Processing: An Introduction to Natural Language Processing, Computational Linguistics, and Speech Recognition*, 2000, Prentice Hall, Upper Saddle River, NJ.

16.16   Glossary

Bag-of-words Condensing a piece of text into its word frequencies.

Entity recognition Using text to identify the specific real-world entities being discussed.

Knowledge base A database storing facts.

Lemma The base, uninflected form of a word.

Ontology A collection of concepts and relationships between them that are required for understanding a particular domain of application.

Part of speech A grammatical part of speech (such as noun, verb, etc.) of a word in a sentence.

POS Short for part of speech.

Sentiment analysis Automatically gauging the tone of a piece of text, especially whether it is positive or negative.

Stem The base part of a word that doesn't change between various forms of the word. Can often be used in place of lemmas.

Stop word A word that is common and not helpful in assessing a text's meaning. In many applications, they are often filtered out because they act as noise.

Topic modeling Identifying topics of discussion in documents. Often, a topic is modeled as a collection of words (such as "football" and "quarterback"), which are usually rare, but which sometimes are all frequent in a document.

TF-DIF A method of weighting the importance of words so that the less common words are more important.

Tokenization Breaking a piece of text into its "tokens." The tokens are usually words but will also often split something like "it's" into two tokens.

17

Time Series Analysis

Time series analysis, in my experience, is not as common as you might expect in data science work. However, that seems to be largely an artifact of the datasets that it has been applied to thus far, which tends to be legacy business spreadsheets and dumps of SQL databases. Especially as sensor nets become more ubiquitous, time series will come to play a much larger role in daily work. At that point, data scientists will have a lot of catching up to do, because electrical engineers have been analyzing time series for decades.

Typical applications of time series analysis in data science include the following:

- Predicting when/whether an event will occur, such as a failure of the machine generating the data
- Projecting the value of the time series at future points in time, such as a stock whose price we want to predict
- Identifying interesting patterns in a corpus of time series data that is too large for a human to comb through.

All of these business applications can ultimately be formulated as machine learning problems. For example:

- If we are trying to predict whether a component is at risk of failure, this is a classification problem: we extract various features from the data to date (especially its recent history) and use it to predict a binary variable of whether it will fail soon (say, in the next hour).
- Let's say we want to predict the value of a time series in the future. Well, finding the value of the time series an hour from now based on recent measurements can be formulated as a regression problem.
- If you are just looking for interesting patterns, this is often accomplished using clustering or dimensionality-reduction algorithms.

Most of this chapter boils down to techniques for converting time series analysis problems into machine learning problems and some of the unique challenges this poses.

The Data Science Handbook, First Edition. Field Cady.
© 2017 John Wiley & Sons, Inc. Published 2017 by John Wiley & Sons, Inc.

17.1 Example: Predicting Wikipedia Page Views

The following example script downloads a time series of Wikipedia page views by day. It then does several things:

- I plot the raw time series and look at it.
- It is visually clear that there are some major outliers, which are likely to confound any model I care to fit, so I cap them off by setting everything above the 95th percentile to the 95th percentile value. This is crude and brute force, but it works for the time being.
- I expect that there will be a weekly periodicity to this signal, so I use the statsmodel library to break it down into a periodic component, a trend component, and a noise component. This is a very error-prone step, but it can be very useful too.
- I break the time series into week-long sliding windows and pull out some salient features from each window. Those features are then used to train a regression model for predicting traffic in the next week.

It's not fancy, but this is the kind of thing you are likely to do at least as a first cut with time series data.

```python
import urllib, json, pandas as pd, numpy as np, \
    sklearn.linear_model, statsmodels.api as sm \
    matplotlib.pyplot as plt

START_DATE = "20131010"
END_DATE = "20161012"
WINDOW_SIZE = 7
TOPIC = "Cat"
URL_TEMPLATE = ("https://wikimedia.org/api/rest_v1"
    "/metrics/pageviews/per-article"
    "/en.wikipedia/all-access/"
    "allagents/%s/daily/%s/%s")

def get_time_series(topic, start, end):
    url = URL_TEMPLATE % (topic, start, end)
    json_data = urllib.urlopen(url).read()
    data = json.loads(json_data)
    times = [rec['timestamp']
        for rec in data['items']]
    values = [rec['views'] for rec in data['items']]
    times_formatted = pd.Series(times).map(
        lambda x: x[:4]+'-'+x[4:6]+'-'+x[6:8])
    time_index = times_formatted.astype('datetime64')
```

```
    return pd.DataFrame(
        {'views': values}, index=time_index)
def line_slope(ss):
    X=np.arange(len(ss)).reshape((len(ss),1))
    linear.fit(X, ss)
    return linear.coef_

# LinearRegression object will be
# re-used several times
linear = sklearn.linear_model.LinearRegression()

df = get_time_series(TOPIC, START_DATE, END_DATE)

# Visualize the raw time series
df['views'].plot()
plt.title("Page Views by Day")
plt.show()

# Blunt-force way to remove outliers
max_views = df['views'].quantile(0.95)
df.views[df.views > max_views] = max_views

# Visualize decomposition
decomp = sm.tsa.seasonal_decompose(df['views'].values,
freq=7)
decomp.plot()
plt.suptitle("Page Views Decomposition")
plt.show()

# For each day, add features from previous week
df['mean_1week'] = pd.rolling_mean(
    df['views'], WINDOW_SIZE)
df['max_1week'] = pd.rolling_max(
    df['views'], WINDOW_SIZE)
df['min_1week'] = pd.rolling_min(
    df['views'], WINDOW_SIZE)
df['slope'] = pd.rolling_apply(
    df['views'], WINDOW_SIZE, line_slope)
df['total_views_week'] = pd.rolling_sum(
    df['views'], WINDOW_SIZE)
df['day_of_week'] = df.index.astype(int) % 7
day_of_week_cols = pd.get_dummies(df['day_of_week'])
df = pd.concat([df, day_of_week_cols], axis=1)

# Make target variable that we
# want to predict: views NEXT week.
```

```
# Must pad w NANs so dates line up
df['total_views_next_week'] = list(df['total_views_
week'][WINDOW_SIZE:]) + \
    [np.nan for _ in range(WINDOW_SIZE)]

INDEP_VARS = ['mean_1week', 'max_1week',
    'min_1week', 'slope'] + range(6)
DEP_VAR = 'total_views_next_week'

n_records = df.dropna().shape[0]
test_data = df.dropna()[:n_records/2]
train_data = df.dropna()[n_records/2:]

linear.fit(
    train_data[INDEP_VARS], train_data[DEP_VAR])

test_preds_array = linear.predict(
    test_data[INDEP_VARS])

test_preds = pd.Series(
    test_preds_array, index=test_data.index)
print "Corr on test data:", \
    test_data[DEP_VAR].corr(test_preds)
```

The script will produce the following outputs:

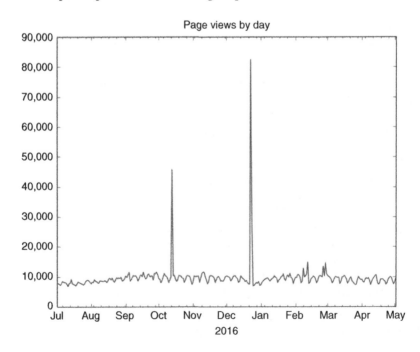

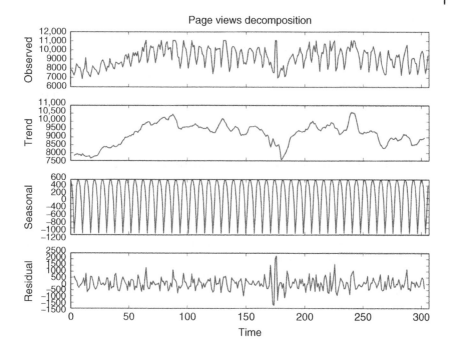

Page views decomposition

```
Corr on test data: 0.78864314787
```

17.2   A Typical Workflow

A typical time series work flow will look something like this:

- Resampling and Interpolation. Time series data often has missing values and/or is sampled at nonuniform rates. However, most algorithms require uniform sampling with no missing data. The first step is to convert the raw input data into uniformly sampled time series.
- Sometimes I'm working with time-stamped events, rather than time series measurements. In this case, I have to first condense these events into time series data. For example, maybe by counting the events per day.
- Series-level preprocessing and denoising. Oftentimes, we want to try various methods of removing noise from the data, smoothing outliers, scaling it to appropriate levels, or other types of preprocessing.
- Windowing. Most applications involve breaking a whole time series down into smaller windows of time, from which we can extract features. How large these windows should be, whether and how much they should overlap, and how we should decide where to place them are all important questions.

- Feature extraction. Once we've broken the series down into windows, we generally want to extract meaningful features from each window that we can plug into a machine learning model.

The other thing that we sometimes do, which is superficially similar to projecting its value but ultimately quite different, is constructing a model of the time series that describes how it behaves as a random process.

17.3   Time Series versus Time-Stamped Events

There are two very different types of data that are sometimes called "time series":

1) Actual time series, that is, a sequence of numbers that are associated with different points in time. This will often be things such as the price of a stock at different points in the day, the amount of revenue that was made each day in a month, or temperature measurements coming off of a physical sensor.
2) Discrete events (or periods of time) that have a time stamp associated with them.

This chapter will focus almost exclusively on the first of these, for two reasons:

1) Most time series analysis techniques are designed to handle numerical measurement data. There's not a lot you can say about event data that generalizes beyond a particular domain of application.
2) Even when we do have event data, often what we really want is still more of a continuous thing. Take Internet traffic, for example. We might care about predicting how many people will visit a website on Tuesday, but we are rarely concerned with how many milliseconds it will be until the next person arrives or how likely it is that there will be five people on the site at any moment in time. Intuitively, there is a continuous-valued "traffic density" that varies across time, and individual human visits are just samples of that density.

Generally, people convert event logs into time series by dividing up time into fixed-size windows and counting the events in each window. Usually, this is done in human-relatable terms such as events per day, events per hour, or something similar, but you can also adjust the windows' size to the granularity that works best for your analyses.

For the rest of this chapter, unless otherwise specified, you can assume that I mean a time series in the traditional sense.

17.4 Resampling an Interpolation

For some terminology, let's say that we have time stamps $t_1, t_2, ..., t_m$, and our measurement at time t_i is $f(t_i)$. In general, the t_i might have different distances between them, but pretty much every algorithm requires equal spacing. So, you need to find a new set of points $T_1, T_2, ..., T_N$ to estimate your signal at, where $T_{i+1} = T_i + \delta$, and get estimates of $f(T_i)$. Oftentimes, the easiest way to do it is to set $T_1 = t_1$, and then have delta be a parameter that you fiddle with.

There are a couple ways to get $f(T_i)$. The easiest is just to find the t_j that is closest to T_i and adopt its value, so that your interpolated $f(x)$ is a piecewise-constant function, as follows:

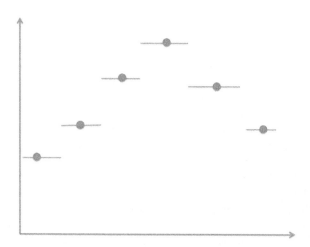

Even easier to implement is backfilling or forward filling, where you just carry $f(t_i)$ forward until you hit the next t_j. These methods are crude, but they are built-in to most libraries you're likely to use, and they're trivially easy to implement yourself if need be. If your t_i are fairly dense, you often don't need anything more. Honestly, it's my own go-to interpolation method, at least as a first cut, since it's trivially easy in Pandas:

```
>>> # Make sure index is already a timestamp
>>> df_indexed = df.set_index('timestamp')
>>> df_sampled = df_indexed.asfreq(
      '1min', method='backfill')
```

The next simplest interpolation method is piecewise linear, which is shown as follows:

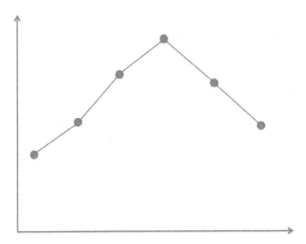

The next step up in sophistication is spline interpolation. Linear interpolation has some obvious pathologies near the known points – surely no real signal will have "teeth" like that. The idea of spline interpolation is that instead of fitting a line to two points, we fit a polynomial to three or more of them. The most common choice is a cubic spline, where values between t_i and t_{i+1} will be interpolated based on t_{i-1} to t_{i+2}. This gives up a nice smooth-looking, continuously differentiable function like we see in the following.

Unfortunately, linear and spline interpolations are not supported in Pandas for interpolating at arbitrary points (although Pandas does let you use interpolation to fill in missing data, with the interpolate() method on Series objects). However, SciPy has a nifty method called interp1d that takes in your data points and returns a callable object that you can treat as an interpolated function. By default, it does linear interpolation, but it can also do cubic or a variety of others if you change the optional arguments. For example:

```
>>> import scipy.interpolate as si
>>> s = pd.Series([0, 2, 4, 6])
>>> s_sqrd = s * s
>>> linear_interp = si.interp1d(s, s_sqrd)
>>> linear_interp([3,5])
array([ 10.,   26.])
>>> cubic_interp = si.interp1d(s, s_sqrd, kind="cubic")
>>> cubic_interp([3,5])
array([  9.,   25.])
```

A major problem in any interpolation scheme is how to deal with interpolating x-values that fall outside of the range for which we have data. Generally, this is called "extrapolation" rather than interpolation, since the point is outside the range of x-values in our data. Extrapolation is a dicey business: the obvious way to do it is to go back to our linear (or other) interpolation, look at the two points closer to the one we want, and extend the line out. But this is likely to give values that are grossly incorrect when you look at the data as a whole. For example:

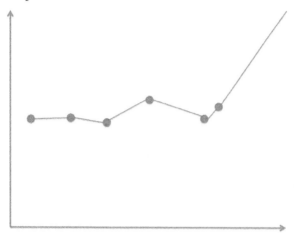

Interp1d will throw an error by default unless you pass bounds_error=False in as an optional parameter. Even then, it will give you a null value for those points rather than an actual extrapolation.

17.5   Smoothing Signals

The simplest way to smooth data is to take the moving average. This just means that we replace each measurement with the average of a fixed number k of measurements before it. Or equivalently, we could replace it with the average of some before and some after it – the difference is just in the time stamp.

It's also common to not just take the normal average, but instead a weighted one. It makes a lot of sense that $f^{smooth}(t_i)$ would be more influenced by $f(t_i)$ than $f(t_{i-1})$.

One of the neatest variations on moving average is called exponential smoothing. In exponential smoothing, $f^{smooth}(t_i)$ will be a weighted average over $f(t_i)$, $f(t_{i-1})$, and so on, with exponentially decreasing weights. This works out the simple formula

$$f^{smooth}(t_1) = f(t_1)$$

$$f^{smooth}\left(t_{i+i}\right) = \alpha f\left(t_{i+1}\right) + \left(1-\alpha\right) f^{smooth}\left(t_i\right)$$

The moving average is great and extremely popular. However, it can suffer if your data contains aberrantly large spikes of noise. In many cases, a massive measurement is a pathology that should be replaced, but the moving average will instead smear the short, tall spike into a slightly wider and flatter one, possibly even making it nonobvious that the original data point was an error. For this reason, I'm a fan of the moving median, where you replace each point in the time series with the median of the k points before it. This is much more computationally intensive than a moving average, but it's more robust. You can also do what I did in the example script at the start of this chapter: cap the values to remove gross outliers.

17.6 Logarithms and Other Transformations

In many domains, it makes sense to apply time series analysis not to the raw data itself, but to some mathematical function of the data. The most important of these is the logarithmic transformation in domains such as finance. There are two reasons you might want to do this:

- In the interest of making your time series meaningful, it is sometimes good to have a movement on the y-axis be equally impactful no matter where it starts. Taking the logarithm accomplishes this.
- Oftentimes, if you try to fit a regression model to your data, the standard models won't do, but they will perform well on the transformation.

17.7 Trends and Periodicity

In the example script, we said

```
decomp = sm.tsa.seasonal_decompose(
    df['views'].values, freq=7)
decomp.plot()
```

The idea here is that we have data that has a strong cyclic behavior (with a periodicity that we know) overlaid with an overall smooth trend component, plus some random noise. By looking at the trend in isolation, we can get a smoother, seasonality-adjusted version of the time series. If the time series was measuring traffic to a website, for example, we could compare Saturday's value with Wednesday's directly, without worrying about the fact that one occurs on a weekend, if we look only at the trend component.

If this sounds too good to be true, then yes: often is. First off, seasonal decomposition only works if the periodicity is exact. That's fine with daily web traffic, but won't work for many processes in the real world where sometimes things take more, sometimes less time. This decomposition also assumes that the noise, seasonality, and trend are additive; it models fixed-size changes in the amount of traffic, but not proportional changes.

The seasonal component is calculated very simply: just by averaging over the corresponding days for the entire time series. The trend curve is then calculated by a fairly complicated process called LOESS. Basically, it fits local polynomials to the data and then splices them together: think like spline interpolation, but without having to cross through all of the data points.

17.8 Windowing

Now we will start to transition from processing that is specific to time series and start looking at how we can turn time series into machine learning problems. In typical machine learning problem, you have clear "entities" that are your objects of study, such as ads that did or didn't get clicked on or human customers in a database. This is not the case with time series data. We just have a list of measurements that were sampled at some frequency, and how you divide that up into "entities" is a touchy question.

The typical approach is windowing, where we select equal-size windows from the signal and separately classify them and extract features. For example, we might

- break our data into 1-hour windows, excluding windows in which the machine we're monitoring failed;
- for each window, extract some features such as average value, Fourier components, and so on;
- label each window according to whether the machine failed within the next hour;
- divide into testing/training data and break out your favorite ML classifier.

Windowing makes a lot of sense when you think of it in terms of common applications. Often in the real world, you want to make a prediction or decision based on the time series data available to date, as of some moment in time. Windowing corresponds to making a decision based on data available as of the end of the window.

If your window length is W and you have N measurement in your time series, then there are $N - W + 1$ possible windows you could select. In the script at the beginning of this chapter, we used all of these windows, but generally, you don't want to do that for two reasons:

1) There are potentially a LOT of data points.
2) The windows will overlap a lot. This is begging for overfitting, and it also means that you are needlessly feeding almost-duplicate data into whatever ML algorithms you end up using.

An alternative option is to line the windows up one after another, so that a new one starts as soon as one ends. This way they are not overlapping, and you are still using all the data. This is often your best option, especially if the window length is "natural" for your domain of application, such as 24-hour periods. You could call this the "covering windows" approach.

There are two potential problems with covering windows:

- If there is periodicity that is the length of the window, then all the windows will look artificially similar. Say, for example, you're measuring something (such as Internet traffic, temperature, etc.) that moves on a daily cycle, and your classifier is trained on 24-hour windows that go from midnight to midnight. If it's 3 pm right now and you're trying to predict whether a component will fail, your data for the last 24 hours will look nothing like any of the training data.
- Often, there are events that you want to line up in the windows in a specific way, and you lose this ability. For example, if you're building an alarm system to predict machine failures, then you probably have several points in time where a failure is known to have occurred. You want to make sure that you have windows in your training data that end shortly before those failures, but with long enough time that you could work to avoid the failure. A failure event is wasted if you put it right in the middle of one of your windows.

However you select your windows, bear in mind that it will affect the statistical validity of your results.

17.9 Brainstorming Simple Features

In no particular order, here are some methods that I have used or considered using for extracting features from time series data:

- The average, median, and quartile values
- The standard deviation of values
- Fit a line to the data and give its slope and intercept. You can fit the line using least squares, an L1 penalty, or any other option.
- Fit an exponential decay/growth curve to the data and use the fitted parameters.

Bear in mind that there is also more to life than windowing. In real situations, you have access to far more data than just the most recent time window, and it makes sense to extract some features that look further back in time. This is especially the case if you have data from many different time series: the same sliding window might mean something very different if the overall time series has been decaying since it started versus staying relatively constant. What I generally do is that for a particular moment in time, I extract some features that are based only on the trailing window and others that are based on the whole lifetime of the series. Some of the latter include the following:

- How long it has been since the time series began, which will often correspond to the age of some physical device.
- The point in time that the series began, expressed as a date. This might be important if, say, devices were set up differently at different points in time.
- The usual aggregate window statistics but applied to a window at the beginning of the time series. This allows for interesting comparisons to the current state of affairs.
- Fit a curve to the whole time series and use the fitted parameters.

17.10 Better Features: Time Series as Vectors

A time series window is just an array of floating numbers. As such it can be treated as a numerical vector like you're familiar with from machine learning. This opens up a wide world of additional techniques.

Most simply, if there is some reference window (perhaps an interesting pattern you've found), you can measure the distance from other windows to the reference pattern. You can use either the normal Euclidean metric

$$\text{dist}(x,y) = \sqrt{\sum_{i=1}^{d}(x_i - y_i)^2}$$

or the so-called taxicab metric

$$\text{dist}(x,y) = \sqrt{\sum_{i=1}^{d} abs(x_i - y_i)}$$

or any other metric of interest to you. If you are only interested in the shape, then normalize the values in each window before calculating. Outside of typical machine learning problems, this approach is also used to find occurrences of a key pattern in a time series (or collection of many time series).

You can also plug your training windows into a clustering algorithm such as k-means. Then, when it comes time to extract features from another window, one of the features can be the cluster of which it would have been a part.

What you will see the most often as a feature, though, is the dot product between your window and one or more reference windows. You might, for example, run PCA on your training windows. The dominant principal components will then represent the major patterns present in your training windows. When it comes time for feature extraction, you can see how much each of each major component is in your window.

Note though that all of these approaches hinge critically on good selection of your windows. If a critical pattern occurs in the beginning of some windows and the end of others, then clustering algorithms won't recognize that they're the same thing. PCA will have problems too. None of these techniques will be rendered completely ineffective, but they will be massively hampered. This is part of why it is so useful if you have a set of points in time where events happened that you know are relevant, and you can pick some of your windows to be located relative to them.

17.11   Fourier Analysis: Sometimes a Magic Bullet

One of the most important techniques in time series is called Fourier analysis; it is often useful in data science and centrally important in engineering and physical sciences. The idea is to decompose your entire signal into a linear combination of signals that vary periodically. For example, we might say that the hourly temperature measured across a decade would have a slow 1-year period (for the seasons) and a fast 24-hour period for the day/night cycle. Fourier analysis is a deep, conceptually rich field of mathematics. In this section, I'm just going to simplify it down dramatically and only discuss the parts of it that are relevant for simple applications in the daily work of a data scientist.

The key theorem is this. Let's say you have an array of N numbers $x_1, x_2, \ldots, x_N$. Then there exist numbers $a_0, a_2, \ldots, a_{N-1}$ and $b_1, b_2, \ldots, b_{N-1}$ such that

$$x_t = a_0 + \frac{1}{N}\sum_{m=1}^{N-1} a_m \cos\left(\frac{2\pi m}{N}t\right) + \frac{1}{N}\sum_{m=1}^{N-1} b_m \sin\left(\frac{2\pi m}{N}t\right)$$

In this expression, a_0 is a constant offset, and every other term is a sinusoidal wave that oscillates with some frequency. Collectively, the a_m and b_m are called the "Fourier coefficients" or, sometimes, the "spectrum" of your signal. The heart of Fourier analysis is a collection of algorithms called Fourier transforms, which let us convert from the raw signal to the spectrum and vice versa.

An alternative version of the Fourier decomposition is to say that

$$x_t = c_0 + \frac{1}{N}\sum_{m=1}^{N-1} c_m \sin\left(\frac{2\pi m}{N}t + \varphi_m\right)$$

In the previous version, for a given m, we had two terms: $a_m \cos\left(\dfrac{2\pi m}{N}t\right)$ and $b_m \sin\left(\dfrac{2\pi m}{N}t\right)$. These are two different sinusoidal signals with the same frequency, which we add together. But a mathematical fact is that there are c_m and φ_m such that they add up to $c_m \sin\left(\dfrac{2\pi m}{N}t + \varphi_m\right)$. This is a single sinusoidal signal, but it happens to be offset in time by φ_m. You can get c_m by just taking

$$c_m = \sqrt{a_m^2 + b_m^2}$$

In practice, you will usually get the a_m and b_m out of a Fourier transform and then switch over to the c_m for practical applications.

If the x array is just a bunch of noise, then the spectrum won't be particularly interesting. But if the signal is periodic or fluctuating, or even approximately so, then it will become glaringly obvious from looking at the spectrum. Often, most of the Fourier coefficients will be small, but a handful will be quite large: in these cases, those few coefficients would allow a pretty good reconstruction of our entire original signal. Often in real systems, especially physical ones, oscillating signals of different frequency correspond to different underlying physical phenomena, so the Fourier decomposition amounts to expressing the data as a linear combination of several real-world processes. The degree to which a physical process is present is indicated by the Fourier coefficients that correspond to its frequency. If we were to, say, use those coefficients as a feature in a machine learning algorithm, then that feature (usually a c_m) will measure a real-world phenomenon.

This might be sounding to you like principal component analysis, where we express our raw signal as linear combination of several "base signals" that are physically meaningful. Yes, you are right. This becomes a linear algebra subject, and both of these approaches fall under the umbrella of finding a "change of basis." But that's an advanced topic; don't worry about it for now.

The algorithm that takes in your raw signal and produces the Fourier coefficients is called the "fast Fourier transform" or FFT. The FFT is one of the most important algorithms in history – the ability to calculate Fourier transforms efficiently was one of the cornerstones of the information age. It comes built into SciPy, as illustrated in the following:

```
>>> from scipy.fftpack import fft, ifft
>>> x = np.array([1.0, 2.0, 1.0, -1.0, 1.5])
>>> spec = fft(x)
```

```
>>> spec
array([ 4.50000000+0.,
        2.08155948-1.65109876j,
       -1.83155948+1.60822041j,
       -1.83155948-1.60822041j,
        2.08155948+1.65109876j])
>>> am = spec.real
>>> bm = spec.imag
>>> cm = np.abs(spec)
>>> x_again = ifft(spec)
>>> x_again
array([ 1.0+0.j,  2.0+0.j,
        1.0+0.j, -1.0+0.j,  1.5+0.j])
```

A note of explanation is due. In this case, x is our raw signal and spec is the Fourier transform. x is an array of length $N = 5$. You might expect spec to be two arrays of numbers, one of length N and another of length $N-1$. Instead, it is a single array of length N, but the numbers it contains are complex numbers (here $j = \sqrt{-1}$ is an imaginary numbers). The way it works is that the mth component of spec will be $a_m + j^* b_m$.

There are theoretical reasons for this: Fourier transforms (unlike many mathematical concepts) work perfectly well with complex numbers, and in fact, they are even more elegant and self-consistent when you think of them with complex numbers. Many electrical engineers, who live and breathe Fourier analysis, make extensive use of the complex nature of these numbers. But for cruder data science applications, we usually turn them back into real numbers as quickly as possible. I don't know about you, but complex numbers hurt my head.

For data science, we typically do the following things with Fourier transforms:

- Using them to identify periodicity in a signal. For example, if you measure blood pressure several times a second, the main frequency in the data will be a person's heart rate.
- Fourier coefficients as features for windows. The amount of, say, 10 Hz frequency in a signal is an incredibly important, perhaps very physically meaningful feature that we can extract and plug into a machine learning algorithm. The one thing is you would have to take the magnitude of the Fourier coefficient rather than the coefficient itself, since it is a complex number.
- Smoothing the data by removing high-frequency jitter. This is sometimes called a "low-pass filter" – you set all the higher coefficients to 0 and reconstruct the signal from that.

- Removing long-time trends to study shorter timescale phenomena. This is called a "high-pass" filter, and it works analogously to a low-pass filter: set the low-frequency coefficients to 0, and then reconstruct the signal.

17.12 Time Series in Context: The Whole Suite of Features

In a typical data science application, multiple, distinct types of data will be brought to bear on a single problem. An example is monitoring machines to predict when they will fail. In situations such as this, you will have continuous-time sensor measurements. But you may also have additional physical specs about each machine, how long it has been running, the conditions under which it operates, and logs of maintenance that have been done on it. It will be your job to extract features out of all this data that can give you predictions for a particular point in time.

So let's assume that you have data of these types, and your goal is to extract meaningful features for making predictions at a point in time T. What can you do?

Most of this chapter has discussed how you can extract features from the time window immediately preceding T. These features are great to have, but they are fairly limited in their scope. Really all they tell us is the state of the machine at time T. But the machine has a whole lifetime that we can look at when making a prediction.

Other features to consider would include the following:

- The age of the machine. It doesn't get much simpler than that.
- How many times repairs have been done in the machine's lifetime.
- How many hours the machine has clocked in a particular state, as measured by doing clustering over all time windows in the training data. The number of hours that a machine has been working hard might be a good proxy for the wear-and-tear on its parts.

As with all areas of data science, the key to using time series is extracting the right features. The key to extracting features is understanding what real-world phenomena are relevant.

17.13 Further Reading

1 Oppenheim, A & Schafer, R, *Digital Signal Processing*, 1975, Pearson, New York, NY.
2 Riley, K, Hobson, M & Bence, S, *Mathematical Methods for Physics and Engineering: A Comprehensive Guide*, 3rd edn, 2006, Cambridge University Press, Cambridge, UK.

17.14 Glossary

Denoising Removing random noise from a time series.

Extrapolation Using interpolation techniques to estimate a function at a point x that is higher than our highest-known x or lower than the lowest-known. This is much more error prone than interpolating at an x that is within our known interval.

Fourier analysis Looking at a time series signal as a linear combination of sinusoids with different frequencies.

Fourier transform The process of going from a raw signal to its Fourier decomposition.

Interpolation Using the known values of a function at several points to estimate its value at a location that we don't know. This is useful when resampling time series data.

Resampling Taking a time series with irregularly spaced time stamps, or time stamps at a frequency we don't want, and estimating the time series at a desired sampling frequency.

Seasonal decomposition Breaking a time series down as the sum of a periodic term, a smooth "trend" term, and random noise.

Spline A method of interpolation where a cubic polynomial is fit to a small number of known points that are close together. That polynomial is used to give interpolated values near those points.

Sliding window Moving a fixed-length window across a time series and calculating some statistic for each window.

Window A contiguous subset of a time series signal. Often, we break a time series into a collection of windows and extract features from each window.

18

Probability

So far, this book has tacitly assumed that you understand basic probability, such as the notion of independence and what an average is. This chapter will go into more detail, giving you a little bit of theoretical background in the subject and an overview of the standard tools. In practice, data scientists only need a moderate amount of probability theory for most of their daily work, but that moderate amount is crucially important. Probability provides the theoretical basis for almost all of machine learning and most of analytics, and it is a critical mindset for data scientists to be able to adopt.

Probability is often confused with statistics. The way I would break it down is to say that probability is a collection of techniques for describing the world using mathematical models that include randomness. In particular, probability focuses on what you can derive about the world assuming that it is well described by one of these models. For example, if we assume a certain distribution of human heights, then how many people in a crowd can we expect to be over 5 ft tall? Statistics is more about working backward: given some real-world data, what can we infer about the real-world process (which we imagine to be some probability model) that generated it?

This chapter will attempt to build up the subject of probability in a very intuitive way. I will start off by showing two of the simplest, most intuitive, and most important probability models. Using these as motivation, I will then zoom out and give a more formal treatment of probability concepts. A certain amount of this will be material that we've already covered in the book; I'll just be discussing it in a more mathematical way. Finally, I will move on to several of the most important probability distributions for you to know as a data scientist.

18.1 Flipping Coins: Bernoulli Random Variables

The simplest probabilistic model is just flipping a (possibly biased) coin. Let's say that the probability of getting a head is p, and hence, the probability of tails is $1-p$. In probability terms, we would say that the flip of such a coin is a

The Data Science Handbook, First Edition. Field Cady.

"Bernoulli random variable" or Bernoulli RV. You might see it denoted as Bernoulli(p).

If p is 0.7, then you can visualize the RV with a bar chart:

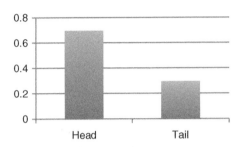

The assignment of 0.7 to heads and 0.3 to tails is called the "probability mass function" for this particular random variable.

It becomes convenient to describe the RV in terms of numbers rather than sides of a coin. The convention here is to say that heads = 1 and tails = 0.

In many cases, there is some kind of payout associated with different outcomes of the random variable. For example, I might give you $5 for every head and demand $2 from you for every tail. The average payout will then be

$$E[\text{payout}] = 0.7 * 5 + 0.3 * (-2) = 2.9$$

The right way to interpret this number is that if you flip the coin N times, where N is some very large number, you will make about $2.9*N$ dollars.

You can see immediately how a Bernoulli random variable might generalize to something such as the roll of a dice, where the probability mass function would assign a probability to the numbers 0–5. In cases such as this, it is conventional to let p_i denote the probability of the ith outcome. The only constraints are that

- all the p_i are nonnegative, and
- they add up to 1.0.

Any set of the p_i that meets these criteria is a valid probability mass function.

A Bernoulli random variable is called a "discrete" random variable. This means that either it has a finite number of outcomes or all of its possible outcomes can be listed out. So, a random variable that assigns a probability mass to every positive integer is still discrete. A random variable that measures human height (to a precision of arbitrarily many decimal places) is not discrete.

18.2 Throwing Darts: Uniform Random Variables

Bernoulli random variables are the simplest type of discrete random variable. The opposite is what are called "continuous" random variables. They can take on any value within a range of numbers.

The simplest continuous random variable is the uniform random variable, sometimes called Uniform(a,b). A Uniform(a,b) will always be between the numbers a and b, but it is equally likely to be anywhere in that range.

For discrete RVs, the probability mass function assigns a finite probability to every possible outcome. For continuous RVs, every exact outcome has probability 0, but certain ranges of outcomes are much more likely than others. We call this relative likelihood the "probability density function" (pdf). The pdf for a uniform distribution appears as follows:

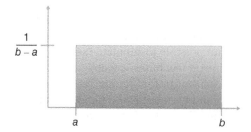

Similarly to probability mass functions, the constraints on a PDF f are that

- $f(x)$ is never negative, and
- the total area under the curve of f is equal to 1.0.

Any function f that meets these criteria is a valid PDF.

Related to the PDF is the "cumulative distribution function" (CDF). Conventionally, if we use the lowercase $f()$ to denote the PDF, we use the uppercase $F()$ to denote the CDF. $F(x)$ is the probability that a random variable's value will be $\leq x$. So, $F(x)$ is a nondecreasing function that goes to 0 as x approaches negative infinity and approaches 1.0 as x becomes large. In places where $f(x)$ is large, $F(x)$ will slope up sharply. In places where $f(x)$ is 0, $F(x)$ will be flat.

The PDF tends to be a lot easier to think about and visualize. However, there are situations where it is easier to solve problems using the CDF.

18.3 The Uniform Distribution and Pseudorandom Numbers

The Uniform(0,1) distribution is the most fundamental probability distribution to understand. It is the simplest one, but it is also the basis for building up many more complicated ones, both in mathematical theory and computational practice. For example:

- If you want to simulate a Bernoulli(p) random variable B, then you can do it by simulating a random value u from a Uniform(0,1) distribution. If $u < p$, then set B = heads. Otherwise, set B = tails.
- If you want to simulate the roll of a weighted dice, divide the range [0.0, 1.0] up into six regions, the ith of which has size equal to the probability of the ith face. Then again draw a value u from a Uniform(0,1) distribution. Have the dice roll be the region of [0.0, 1.0] into which u falls.
- If you want to simulate an exponential random variable (to be discussed later), draw u from a Uniform(0,1). Then take −1 times log(u).

In general, say you know the CDF $F_X()$ of a random variable X. Say also that you're able to compute the inverse of $F_X^{-1}(u)$. Then $F_X^{-1}(u)$ will be a sample of X if u is drawn from a Uniform(0,1). For these reasons, computational libraries that simulate random variables tend to start with sampling the uniform distribution as their most fundamental operations and build everything up from there.

Now technically, it is not possible to simulate random numbers with a computer. They are deterministic machines, capable only of following predetermined rules – there is no subroutine for flipping a coin. So the standard practice is instead to use "pseudorandom numbers." The idea is this: you start with an arbitrary sequence of bytes. The bits of that sequence are interpreted as the digits of a Uniform(0,1) random number, expressed in binary out to a fixed number of decimal places. In some implementations, only part of the array is interpreted as a number. There is then a complicated (but deterministic!) mathematical function that mangles the byte array into a new byte array. The new byte array is technically a deterministic function of the old one, but in practice, it bears no noticeable resemblance to the original array. Flipping a single bit in the original byte array could change bits anywhere in the output. The new array is treated as a new Uniform(0,1) variable, and so on.

A lot of work has gone into creating pseudorandom numbers that accurately ape all the properties of true randomness. Technically, each sample is a deterministic function of the sample before it, but there is no noticeable correlation between them. If you draw enough samples, they will occur equally often in all parts of the range [0.0, 1.0]. They will have the correct average, standard deviation, and so on. In short, if the pseudorandom samples had been given to us as a stream of raw *data* rather than something we had generated ourselves, we would never have figured out that they were anything other than independent samples from a Uniform(0,1).

The one great thing about pseudorandom numbers is that you can manually set the initial byte array at the start of a program. In this case, it is called the "seed." If you do this, then the program becomes fully deterministic, and you can reproduce it exactly between two runs. This means the following:

- If there is a bug in a randomized program that only occurs sometimes, you can make it perfectly reproducible and figure out what's going on.
- If you need your analytics results to be exactly reproducible because somebody will scrutinize them, you can set the seed in your scripts.
- When you are writing tests, you can set the seed and make sure that the output is exactly what's expected.
- Oftentimes, you have two pieces of code that you need to make sure work identically (maybe a proof-of-concept and a production version). The easiest way to do this is to make sure that they produce the same output, given the same input. This becomes impossible if the code includes calls to random numbers, unless you set both of them to have the same random seed.

18.4 Nondiscrete, Noncontinuous Random Variables

Mathematically speaking, you can have random variables that are neither discrete nor continuous. For example, take the heights of trees that you have planted. At a given point in time, a certain fraction of them will not have sprouted and will have height 0. This is a finite probability mass at that number. But of those trees that *have* sprouted, their heights can fall anywhere within a range.

In practice, this is not a big deal. You're not dealing with abstract probability distributions, but finite datasets; a hybrid distribution would show up just as there being multiple identical numbers in the otherwise continuous-valued data. Calculating the mean, average, median, or other metrics of interest would still be exactly the same procedure.

The place you will run into a problem is with exploratory visualizations. If you do a histogram of heights with our tree example, you will see a massive spike at height = 0. The bell-curve part of the histogram will be squashed down to the point where it's invisible. A better visualization includes two pictures: a pie chart showing how many heights are zero and how many are nonzero, plus a histogram of only the nonzero heights.

The following script simulates some data like this and shows the two visualizations:

```
import numpy as np
import pandas as pd
import matplotlib.pyplot as plt
z = np.zeros(1000)
x = np.random.exponential(size=1000)
data = np.concatenate([z, x])
pd.Series(data).hist(bins=100)
plt.title("Huge Spike at Zero")
```

```
D = pd.Series(data)
X = pd.Series(x)
D.hist(bins=100)
plt.title("Huge Spike at Zero")
plt.show()
(D>0).value_counts().rename(
    {True:'> 0', False:'= 0'}).plot(kind='pie')
plt.title('Breakdown by equal/greater than Zero')
plt.show()
X.hist(bins=100)
plt.title("Distribution When > 0")
plt.show()
```

The naïve histogram looks as the following:

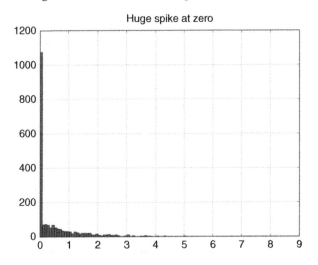

But the pie chart/histogram combo gives us

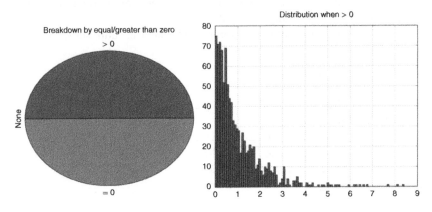

In line with the large spike in the first histogram, you can think of hybrid distributions as having a PDF that has an "infinitely tall spike" at one or more places, and the area under the nonspiked parts of the curve is less than 1. This isn't mathematically rigorous, but it is useful way to think about it. What *is* rigorous though is the cumulative distribution function: the CDF jumps up at every spike.

18.5 Notation, Expectations, and Standard Deviation

Now that you are familiar with some of the key concepts, let's dive into some of the standard notation and terms.

As we've already seen, a "random variable" (RV) is any quantity that can turn out in several different ways. It is typical to use an uppercase letter such as X to refer to the random variable and a lowercase x to refer to a specific value that the variable took on.

A single-dimensional random variable is described by a probability mass function if it is discrete or a probability distribution function if it is continuous.

You can also have a random variable that returns a random vector of d dimensions. The concepts of probability mass function and probability density generalize naturally. The only constraints on them are that they are always nonnegative, and the probabilities add up to 1.0 (or the area under the curve is 1.0 in the case of continuous RVs). Unless otherwise stated, I will typically tacitly assume that RVs in this.

If we write

$$E[X]$$

we mean the average value of the random variable X. The use of "E" in this case refers to "expectation value," which is another term for the mean or average. The expectation value is defined to be

$$E[X] = \sum_i i p_i$$

for discrete RVs and

$$E[X] = \int x f_X(x) dx$$

for continuous RVs. It is common to denote the expectation value of a random variable X by μ_X.

More generally, you can have a function $g()$ of a random variable, and say

$$E[g(X)] = \int g(x) f_X(x) dx$$

and similarly for discrete variables. Many key probability concepts can be defined in terms of expectation values of different functions.

A key example of something being defined in terms of expectation values is the variance and standard deviation. The "variance" of X is defined to be

$$\text{var}[X] = E\left[|X - \mu_X|^2\right]$$

and the standard deviation is its square root

$$\sigma_X = \sqrt{\text{var}[X]}$$

The standard deviation gives you, roughly speaking, a measure of how far X "typically" is from μ_X.

Note that using $E[g(X)]$ lets us use the same notation for discrete or continuous RVs, which can be quite convenient.

18.6   Dependence, Marginal and Conditional Probability

Oftentimes, you have two random variables X and Y and want to consider their behavior together. Does knowing something about one tell you something about the other? For concreteness, let's assume that they are both discrete RVs, and let p_{xy} denote the probability that $X = x$ and $Y = y$.

The "marginal" probability mass function of X is then

$$\Pr[X = x] = p_x = \sum_y p_{xy}$$

and similarly

$$\Pr[Y = y] = p_y = \sum_x p_{xy}$$

These are the probability distributions that we get if we focus on one RV and ignore the other.

On the flip side, if we *know* the value of X and want to infer something about Y, then we care about the *conditional* probability of each y given that $X = x$:

$$\Pr[Y = y \mid X = x] = p_{y|X=x} = \frac{p_{xy}}{\sum_g p_{xg}}$$

Conditional probabilities play a very central, explicit role in Bayesian statistics.

It is common to want to know something such as the expectation value of Y, given that $X = x$. We write this as

$$E[Y \mid X = x]$$

We say that such statistics about Y are "conditioned on" the value of X.
The correlation of X and Y is defined to be

$$\text{Corr}[X,Y] = \frac{E\left[(X - \mu_X)(Y - \mu_Y)\right]}{\sigma_X \sigma_Y}$$

This is a measure of the degree to which there is a linear relationship between the RVs, of the form $Y = mX + b$.

X and Y are "independent" random variables if knowing either tells us nothing about the other. Mathematically, this means that

$$p_{xy} = p_x p_y$$

It's important to note that independence is an extremely strong criterion. It is much stronger than just saying that the correlation is zero.

18.7 Understanding the Tails

One of the most important things to understand about a probability distribution is how "heavy-tailed" it is. Intuitively, this refers to how often you have extremely large values. Human heights are a great example of something this is *not* heavy-tailed: there is never a person who is over 10 feet tall. Net worth, however, is extremely heavy-tailed, since there is the occasional Bill Gates.

Heavy-tailed distributions are extremely common in data science, especially web-based applications. Places that I've seen them arise include web traffic (a few websites are *much* more popular than others) and bids in online auctions.

Heavy-tailed distributions are important to be aware of because many of the usual things we do with probability distributions don't work (at least not in the same way) when things are heavy-tailed.

The average value of a heavy-tailed distribution can be notoriously hard to estimate. If you have 100 random people in a room, the average net worth in the room could vary widely – there is a very realistic chance of having a multimillionaire who will throw off the average net worth.

To see how this works, the following script will simulate a sequence of samples drawn from a Pareto distribution, which is heavy-tailed. It will then plot N on the x-axis and the average of the first N samples on the y-axis. You can see that over 1000 trials, the average does *not* converge to a nice sane average. Instead, it occasionally jumps up when a really massive outlier is hit, then trails off as the subsequent samples are more modest. As time goes on, it takes larger and larger outliers to cause a large bounce. So, the jumps get rarer and rarer, but they do come. If you ran the simulation forever, the average would increase to infinity.

```python
import numpy as np
import matplotlib.pyplot as plt
np.random.seed(10)
means = []
sum = 0.0
for i in range(1, 1000):
  sum += np.random.pareto(1)
  means.append(sum / i)
plt.plot(means)
plt.title("Average of N samples")
plt.xlabel("N")
```

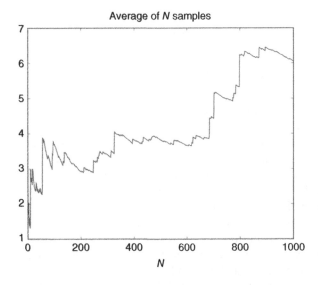

There are a lot of heavy-tailed distributions, and not all of them would display an infinitely growing average. Most real-world ones will grow for a while and then eventually plateau at a finite mean. But it will take a very long time for this to happen.

So calculating the average of a heavy-tailed distribution is a dicey proposition. But even once you have it, the average is useless for many applications. It is not a "typical" value. Most samples are well below the average value; it's just that the few outliers are so large they pull it up.

Metrics such as the median and quantiles are robust to heavy tails. They have the same reasonable interpretations that they always did. However, in many applications, they aren't particularly useful because the large outliers are precisely the values that you are interested in!

So, always keep in mind whether a process you are studying is liable to be heavy-tailed. If it is, make sure that you model it with appropriate distributions and treat it with caution.

18.8 Binomial Distribution

A Binomial(n, p) distribution is the number of heads you get from tossing a coin n times, where each toss has an independent probability p of coming up heads. The key things to understand about it are as follows:

- It can take on any value from 0 to n.
- The average value will be np, and the probability mass function will be peaked there.
- The standard deviation goes as $\sqrt{n}$. This means that as n grows larger, the probability distribution will be more and more strongly peaked around np.

It is instructive to derive the precise formula for Binomial(n, p)'s probability mass function. The key insight is that for any specific sequence of n flips that contains k heads, the probability of tossing that sequence is $p^k(1-p)^{n-k}$. This is because there are k tosses that must come up heads, each with probability p, and $n - k$ tosses that must come up tails, each with probability $1-p$. So to calculate p_k, we only need to ask how many such sequences there are.

How many ways you can get k heads out of n tosses is a combinatorics question that lies outside the context of this book. To motivate it though, imagine that n is very large. Then:

- If $k = 0$, there is only one possible sequence: n tails.
- Similarly, if $k = n$, there is only one sequence: all heads.
- If $k = 1$, there will be n possible places it could be.

The exact formula is denoted $\binom{n}{k}$ and pronounced "n choose k." It is equal to

$$\binom{n}{k} = \frac{n!}{k!(n-k)!}$$

where $x! = x*(x-1)*(x-2)*...*3*2*1$ is pronounced "x factorial."

To sample from a binomial distribution in NumPy, you can write

```
import numpy as np
sample = np.random.binomial(200, 0.3)
```

18.9 Poisson Distribution

Poisson distributions are used to model systems where there are many events that could happen, and all are independent of each other, but on average only a few of them will happen. A good example is how many people will visit a website on a given day; there are billions of people in the world who could visit, but on average perhaps only a few hundred will do it.

Imagine taking a Binomial(n, p) distribution. Set n very large, and set p small enough such that

$$np = \lambda$$

where λ is some fixed constant. In the limit of making n large and p small, but keeping λ fixed, a binomial distribution will converge to a Poisson distribution. The probability mass function is given by

$$p_k = e^{-\lambda}\frac{\lambda^k}{k!}$$

To sample from a Poisson distribution in NumPy, just write

```
sample = np.random.poisson(200)
```

18.10 Normal Distribution

If there is one probability distribution to know, it is the normal distribution, also called the Gaussian. It is the prototypical bell-shaped curve, and its PDF is

$$f(x) = \frac{1}{\sqrt{2\pi\sigma^2}}e^{-(x-\mu)/2\sigma^2}$$

where μ is its average value and σ is the standard deviation. Such a normal distribution is often called $N(\mu, \sigma^2)$. The PDF is displayed in the next page.

The most important practical property of the normal distribution is that its probability density is tightly clustered around the mean; it has very light tails, and major outliers are extremely unlikely. For this reason, it can be dangerous to naively fit a normal distribution to your data. In practice, it is common to identify and remove major outliers before fitting a normal distribution to the remaining data.

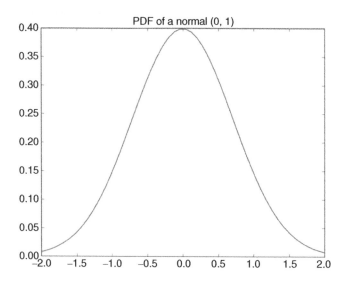

Theoretically, the normal distribution is most famous because many distributions converge to it, if you sample from them enough times and average the results. This applies to the binomial distribution, Poisson distribution and pretty much any other distribution you're likely to encounter (technically, any one for which the mean and standard deviation are finite).

This is captured in the "central limit theorem," which states

> *Central Limit Theorem.* Let X we a random variable with finite mean μ and standard deviation σ. Let $X_1, X_2, ..., X_n$ be a sequence of independent samples of X. Then as n approaches infinity

$$\sqrt{n}\left(\frac{1}{n}\sum X_i - \mu\right) \to N\left(0,\sigma^2\right)$$

18.11   Multivariate Gaussian

Most of the distributions in this chapter are univariate and don't generalize in a noteworthy way to higher dimensions. The normal distribution is an exception to that. A normal distribution can be defined for any number d of dimensions. The density function always resemble a "hill," which peaks at the distribution's mean and is ellipsoidal in shape.

In two dimensions, the following shapes give you some idea of what these ellipsoids can look like the following:

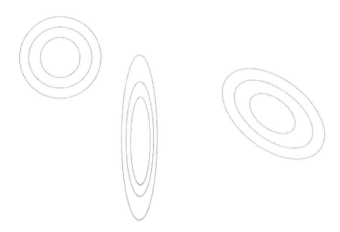

Note that the ellipsoids can stretch out in any direction; they don't have to stretch along one of the axes.

The univariate normal distribution was parameterized by the mean μ and the variance σ^2, both of which are floating-point numbers. For a multivariate Gaussian, the parameters are a d-dimensional vector μ, and a d-by-d matrix Σ. Σ is called the "covariance matrix," and its (i,j)th component will be $\mathrm{Cov}[X_i, X_j]$, where X_i and X_j are the ith and jth components of the random variable.

The PDF at a d-dimensional point x is

$$f(x) = (2\pi)^{-k/2}|\Sigma|^{-1/2}\,e^{-\frac{1}{2}(x-\mu)'\Sigma^{-1}(x-\mu)}$$

The multivariate Gaussian has the same strengths and weaknesses as the univariate Gaussian. It is mathematically convenient, and there are many theorems to the effect that other things will converge to it. On the other hand, it has very thin tails and hence does not allow for large outliers.

18.12 Exponential Distribution

The PDF of an exponentially distributed random variable looks like the following:

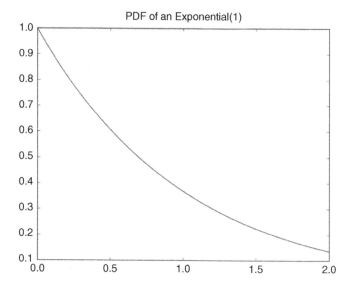

Exponentials show up in a lot of places, but they are most useful in modeling the time until some event occurs or the length of time between events. Let's say, for example, that you have people walking into a store. Every moment there is a small, fixed probability of somebody walking in, and every moment is independent of every other. In that case, the amount of time between events will be exponentially distributed.

The exponential distribution is parameterized by its mean θ, the average time between events. Sometimes, you will instead see it parameterized by $\lambda = 1/\theta$ – the average rate at which events occur.

The PDF of the exponential distribution is

$$f(x) = \begin{cases} \dfrac{1}{\theta} e^{-x/\theta} & \text{if } x \geq 0 \\ 0 & \text{otherwise} \end{cases}$$

To sample from it in NumPy, you can say

```
np.random.exponential(10)
```

In many applications, the key property of the exponential distribution is that it is "memoryless." No matter how long you have been waiting for an event to happen, the time you have left to wait still follows the same exponential distribution. An event will happen in the next moment – or not – independently of what other events happened previously.

The memoryless property of the exponential distribution is usually taken as the dividing line between heavy-tailed distributions and those that are not heavy-tailed. If we have already waited x time for an event to occur, do we expect to wait *more* or *less* time than when we first started? For exponential random variables, it tells you nothing. In contrast, somebody who is 20 years old is likely to like 20 more years, but somebody who is 90 probably won't. Hence, age is not heavy-tailed. A random person on the street is unlikely to be a millionaire. But if you *do* happen to pick somebody who has at least $10 million, there's a decent chance that they have many millions more. Hence, net worth is heavy-tailed.

18.13 Log-Normal Distribution

My go-to heavy-tailed distribution is the log-normal. It is straightforward to understand and simulate. Plus, it always has a finite average and standard deviation, which is true of essentially all real-world phenomena.

Its PDF looks like the following:

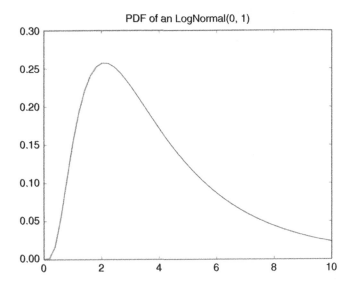

There is a well-defined peak in the distribution, and the peak is at a location above 0. To the left of the peak, it falls off quickly, becoming 0.0 at $x = 0$. To the right though, it tapers off much more gradually, allowing for a regular occurrence of large outliers.

The log-normal distribution is best thought of in this way: sample a value x from a Normal(μ, σ^2). Then e^x is log-normally distributed.

To sample from it in NumPy, you can say

```
np.random.lognormal(mu, sigma)
```

18.14 Entropy

Entropy is a way to measure "how random" a random variable is, and it comes from the field of information theory. Intuitively, a fair coin is more random than one that is heads 99% of the time. Similarly, if a normal distribution has a tiny standard deviation, then its probability mass will be tightly grouped around its mean, and we want to think of it as being less random than one with a larger standard deviation.

Entropy is notoriously difficult to give a crisp explanation of, but I'll do my best. To start off with, let's assume that the random variable X is discrete. The key intuition is that some outcomes of a random variable are "more surprising" than others, and entropy is $E[\text{Surprise}[X]]$. The question then is how to quantify how surprising a certain outcome $X = x$ is. The standard way to do it is that

$$\text{Surprise}[X = x] = -\ln(p_x)$$

In that case, the entropy becomes

$$H[X] \equiv E[\text{Surprise}[X]] = -\sum_x p_x \ln(p_x)$$

If X is a continuous random variable, the definition is

$$H[X] = -\int f(x)\ln(f(x))dx$$

The choice of the logarithm as a measure of "surprise" might seem arbitrary, but it's not. Intuitively, we would like the surprise to be a function of the probability, so that

$$\text{Surprise}[X = x] = f(p_x)$$

where f is some function. We would like this function to have three key properties. The first two are very straightforward:

- $f(1.0) = 0$
- As $p \rightarrow 0$, then $f(p) \rightarrow \infty$

The third property is a little bit more subtle. Say there are two random variables X and Y that are independent of each other. In this case, the surprise from learning about one shouldn't affect our surprise from learning about the other. This means that

- $f\left(p_{X=x}p_{Y=y}\right)=f\left(p_{X=x}\right)+f\left(p_{Y=y}\right)$

Together, these constraints require that you use a logarithm.

Entropy is used in a lot of contexts. Probably the most common is when we are picking a probability distribution to use for some purpose, and we want to pick one that reflects a large amount of ignorance about the situation because we don't know much about it. This will often give us a rigorous criterion for picking one distribution (or a family of distributions) over another. For example:

- If you need a discrete random variable defined over the number 1 to N, then the maximum entropy distribution will assign a probability of $1/N$ to each number.
- If you need a continuous random variable defined over the interval $[a, b]$, then the maximum entropy distribution will be Uniform(a, b).

Whenever you are doing entropy maximization, constraints of some kind are necessary. If an RV was defined over all the real numbers, for example, you could make the entropy arbitrarily large by spreading the probability mass out very thinly over a very wide area.

One final note on entropy: the definitions for discrete and continuous random variables are *not* equivalent. Let's say that you tried to apply the discrete definition to a Uniform(a, b) random variable. You might do so by picking a large number n, dividing $[a, b]$ into n equally-spaced intervals, and looking at the entropy of which interval it falls into. Then the entropy you would get is

$$
\begin{aligned}
H &= -\sum_{i=1}^{n}\frac{1}{n}\ln\left(\frac{1}{n}\right) \\
&= \sum_{i=1}^{n}\frac{1}{n}\ln(n) \\
&= \ln(n)
\end{aligned}
$$

This expression blows up to infinity for large n! The way to think about this is that for a continuous RV, you could never correctly guess the *exact* outcome of a sample, since you would have to be right to infinitely many decimal places. So in that sense, a continuous RV is always infinitely surprising. That's the reason why we use a different definition, one that reflects whether we can make a guess that is *close* to the right one.

You rarely calculate entropy in the course of doing data science work. However, it is ubiquitous as soon as you start getting into the construction and theoretical properties of the tools.

18.15   Further Reading

1 Ross, S, *Introduction to Probability Models*, 9th edn, 2006, Academic Press, Waltham, MA.

2 Feller, W, *An Introduction to Probability Theory and its Applications*, Vol. 1, 3rd edn, 1968, Wiley, Hoboken, NJ.

18.16   Glossary

Bernoulli random variable A random variable describing the flip of a coin that is heads with some probability p.

Binomial random variable A random variable that is the number of heads in n flips of a coin that is heads with probability p.

Central limit theorem A theorem describing how the average of many samples from a probability distribution is normally distributed around the distribution's mean. It applies so long as the distribution being sampled has finite mean and standard deviation.

Continuous random variable A random variable that takes on values in a continuous set, such as the real numbers. Any single number has probability 0 of being the output. Instead, the probability is described by a PDF.

Cumulative distribution function (CDF) If $F()$ is the CDF of a random variable X, then $F(x) = \Pr[X \le x]$.

Discrete random variable A random variable that takes on values in a discrete set, such as {heads, tails} or the integers.

Entropy A measure of how unpredictable a random variable is. For a discrete RV, it is $H[X] = -\sum_i p_i \ln(p_i)$. For a continuous random variable, it is

$$H[X] = -\int f(x)\ln(f(x))\,\mathrm{d}x.$$ These definitions are not equivalent; entropy

is one of the few areas where continuous and discrete RVs require different treatments.

Expectation value The average value of a random variable or of a function of a random variable.

Exponential random variable A type of random variable, which takes on nonnegative numbers. Its probability density peaks at 0 and falls off exponentially. It is often used to model the time between events in a random sequence of events.

Gaussian Another name for the normal distribution.

Heavy-tailed distribution A distribution with heavier tails than an exponential distribution. Practically, this means that large outlier values are more frequent than with an exponential distribution.

Log-normal distribution A specific heavy-tailed distribution. A sample from it is obtained by sampling from a normal distribution and then taking the exponent of that value.

Memoryless A property of the exponential distribution. If you know that $X > x$, then the distribution of $X - x$ is the same as the original distribution of X.

Normal distribution The prototypical "bell-shaped curve." This is a probability distribution with a single clear peak, with distribution that is symmetric on each side of the peak and has very thin tails.

Poisson random variable A random variable defined over the nonnegative integers. It is used to model how many events happen, when infinitely many *can* happen, but in practice, only a few are expected, and those events are independent of each other. Traffic to a website on a given day is often modeled as a Poisson RV.

Probability density function (PDF) A function that describes the probability distribution of a continuous RV. The PDF for a random variable X must be nonnegative, and the total area under it must be 1.0. The area under only a part of the PDF is the probability that X occurs in that region.

Probability mass function The analog of the PDF for discrete random variables. It is a function that takes each possible outcome of the random variable and gives the probability of it occurring.

Pseudorandom numbers A sequence of numbers that are deterministically generated by a computer, but which in practice behave as if they were random.

Random variable A quantity that randomly takes on any of a number of possible values.

Standard deviation A measure of how far a random variable "typically" is from its mean. It is defined as the square root of variance.

Uniform random variable A continuous random that only takes on values within a range $[a, b]$. Within that range, however, the PDF is flat and each area is equally likely.

Variance $\text{Var}[X] = E[(X - E[X])^2]$.

19

Statistics

I should start off with an explanatory note. A lot of data science really should be considered a subset of statistics. It is largely a matter of historical accident that statistics, data science, and machine learning are seen as different things. The disciplines have evolved largely independently, focusing on very different problems, so they have become different enough that I treat them as separate things in this book.

Most data scientists, most of the time, don't really need a thorough knowledge of statistics. There are some who live and breathe it, to be sure, but it's not nearly as useful for data science as one might expect. What's absolutely crucial, however, is the kind of critical thinking that one usually learns in a statistics class. Statistics is all about being extremely, painstakingly careful and rigorous in how we analyze data and the assumptions we make. Data science focuses more on how to extract features out of data, and there is usually enough data available that we don't need to be so exceedingly careful. But data scientists need to be sensitive to the luxury provided by having a lot of data and able to break out more rigorous methods when the data is lacking.

This chapter will cover several of the key topics in statistics. In each case, it will focus on the key ideas, insights, and assumptions underlying each topic, rather than rigorous derivations of each formula.

19.1 Statistics in Perspective

It might seem absurd that most data scientists don't need statistics. They obviously use some statistical tools such as averages and fitting a line, but how can the mother discipline be demoted to a footnote? The way I think about it is this: the discipline of statistics is about how to deal with constraints that data scientists don't usually need to worry about.

The most important of these constraints is sample size. Data science grew out of Big Data, where you are almost by definition swimming in data points.

The Data Science Handbook, First Edition. Field Cady.
© 2017 John Wiley & Sons, Inc. Published 2017 by John Wiley & Sons, Inc.

In situations such as web traffic, there are more data points than you even need and more features than you know what to do with. Your task is to figure out the right way to parse it. Once you've done the leg work of pulling features out of the data, extracting a business insight can be as simple as looking at a histogram.

But this isn't always the case. If you are testing whether a fertilizer works on crops, every data point will require a significant piece of land to be set aside to experiment on and then a year for the crop cycle. If you are testing whether a medical procedure works, every data point you gather is literally a life-and-death proposition. Statistics is about dealing with these extremely constrained situations, where it is very hard to tell whether a pattern we observe is a truth of the world or a fluke of our data. Billions of dollars and potentially even human lives are on the line, so statistics tends to be appropriately nitpicky about every little detail.

Again, data scientists typically have so much data that they can usually afford to be cavalier. But they can't always be, and one of the most serious crimes a data scientist can commit is to be fast-and-loose in a situation that demands a more rigorous approach. The most important thing for you to get out of this chapter is a sensitivity to when quick-and-dirty approaches break down. If you can learn to spot those pitfalls before they happen, you can always learn the statistics you need on the fly.

19.2 Bayesian versus Frequentist: Practical Tradeoffs and Differing Philosophies

You've probably heard of the great divide in statistics between Bayesian statistics and Frequentist (aka classical) statistics. Before getting into the details of the differences, I should let you know that the debate is more philosophical than practical; in most problems, they will give close to the same answers.

Bayesian and Frequentist statistics both take in data and then use it to construct a statistical model of the world (such as a normal distribution). The difference is in the relationship between the available data and the models we construct.

In classical statistics, the model should be a "best fit" to the data. We are obligated to make some assumptions about the form of the model (such as having it be a normal distribution) in order to solve problems, but otherwise, we set the model parameters so as to best fit the data. The overriding paradigm in classical statistics is to ask how plausible our data is given a particular model of the world or particular set of parameters and set our parameters so that the data becomes as plausible as possible. If we must make statistical predictions, we do so using this best-fit model.

In Bayesian statistics, there is an additional layer of complexity. We don't just have the best-fit parameters for the real-world model; we have a confidence distribution over what those best-fit parameters might be. This confidence distribution (which is mathematically equivalent to a probability distribution) isn't a rigorous best fit to any data: it represents our fallible human belief about what the "real" probability distribution might be. A Bayesian model starts off with what's called a "prior," which exists without any data. A prior is the initial confidence distribution over possible models of the world. As data comes in, we refine the confidence distribution, hopefully zeroing in on the "real" parameters that characterize the world as it actually is. If we want to actually make predictions using a Bayesian model, we must average our predictions over all the possible values of a model, weighting them by our confidence.

With any luck, training a Bayesian model will zero in on real-world parameters that are pretty close to the best-fit parameters of Frequentist statistics, and a model with those parameters will be a pretty good model of the world.

Classical and Bayesian statistics both have their place. Bayesian is especially useful in situations where there is expert knowledge available that can be wrapped into a prior, or we have a lot of missing data. Classical statistics is often much easier to compute and make use of. This chapter will start with classical techniques and then move on to Bayesian statistics at the end.

19.3   Hypothesis Testing: Key Idea and Example

An important domain of statistics is called hypothesis testing. The basic idea is that you think that there is a trend in your data, and you want to assess how likely it is to be a real-world phenomenon versus just a fluke. Hypothesis testing doesn't address the question of how strong a trend is; it's designed for situations where the data is too sparse to quantify trends reliably.

A prototypical example for hypothesis testing is this: I have a coin that might or might not be fair. I throw it 10 times and get 9 heads. How likely is it that this coin is loaded?

The first thing to realize is that we can't answer the question "how likely is this coin to be loaded?" In order to get that number, we would need some a priori knowledge or guess about how likely the coin is to be loaded, plus how loaded it would actually be, and there's no principled way to acquire that.

So, here is the first big idea in classical statistics: instead of saying how likely the dice is to be loaded given the data, we ask how likely the data is *given* that the coin is not loaded. Basically, is it plausible that nine heads is just a fluke? We assume a fair coin, compute how likely this skewed data is, and use that to gauge whether the fair coin is believable.

So, let's assume a fair coin. Every coin flip can be either heads or tails, and they are all independent. This means that there are $2^{10} = 1024$ possible ways that the sequence of 10 flips could occur, and they're all equally likely. There are four ways we could get data that is as skewed (or more skewed) as what we observed:

- We could get 10 heads. There is only one flip sequence that does this.
- We could get 9 heads. The tail could be in any of the flips, so there are 10 ways to get 9 heads.
- We could get 10 tails. There is only one flip sequence for this.
- We could get 9 tails. There are 10 ways to get this.

Adding all of these up, there are 22 ways to get data that is as skewed as what we saw. Then we have $22/1024 = 0.0215$, so there is only a 2% chance of getting data such as this from a fair coin. Personally, I don't trust the coin.

That's the basic procedure, and I encourage you to keep it firmly in mind. Now let's frame it in more general statistical parlance. What we have done is called "hypothesis testing": you have found a pattern in the data, and you quantify your confidence in it by calculating how likely that pattern is to be just a fluke.

The idea of a fair coin is called the "null hypothesis". Typically, the null hypothesis means that there is no pattern in the real world: the fertilizer doesn't do anything for crops, the medicine doesn't help or hurt patients, the coin is fair, and the dice is not loaded. Framing the null hypothesis can be complicated for some problems, but it's dead simple in this case.

Then there is some "test statistic": a single number that we calculate from our data that quantifies the pattern we are looking for. The key thing about the test statistic is that we can calculate its probability distribution if we assume the null hypothesis. In this case, the test statistic is just the number of heads. In general though, picking a good test statistic can be a tricky problem.

Finally, there is the likelihood of the test statistic being as extreme as we have observed it being. Basically, the likelihood of seeing a pattern that is this extreme in the data if we take the null hypothesis as a given. This number is called the "p-value." A result is called "statistically significant" if the p-value is below some specified threshold, and there is a widespread convention of using 0.05 as the cutoff. I can't emphasize enough though that this is arbitrary, and we're really dealing with a sliding scale: a p-value of 0.01 is much more meaningful than 0.04, and conversely p-value of 0.07 could still be very important.

The most important caveat to hypothesis testing is that it tells us that a pattern exists, but does not tell us how strong the pattern actually is. It's designed for situations where there is so little data that this knowledge is all we can hope for. In the real world, the null hypothesis is rarely 100% true; the weight of Abraham Lincoln's copper nose will technically introduce a tiny bias in a coin flip, and if you toss the coin a million times, you will see it. I like to say that in Big Data situations, if your pattern is weak enough that you even have to ask the p-value, then it is certainly too weak to be of any business value.

The other important caveat is that you will get false positives. Say you do many hypothesis tests in your life, and you always use 5% as your threshold. Then of the times when there really is no pattern, 1 in 20 of those times you will find something statistically significant. This phenomenon is a huge problem is the scientific literature, where researchers will often declare victory as soon as the *p*-value dips low enough to publish a paper.

19.4 Multiple Hypothesis Testing

In real situations, we often have several hypotheses that we want to test and see if any one of them is correct. For example, we might test 20 different ads out on different groups of users and see if any one of them increases the click-through rate (relative to a known baseline click-through probability) with confidence threshold 0.05. That is, we will calculate the *p*-value for each ad and consider all those with *p*-value < 0.05 to pass our test.

The problem is that while each individual ad has only a 5% chance of passing our test by dumb luck, there is a very good chance that *some* ad will pass by dumb luck. So, the idea is that we must tighten *p*-value constraints for the individual ads, in order to have a *p*-value of 0.05 for the overall test.

A standard, conservative, and surprisingly simple solution to this problem is called the Bonferroni correction. Say you have n different ads, and you want a *p*-value of α for your overall test. Then you require that an individual ad passes with the smaller *p*-value of α/n. We then see that if the null hypothesis is true (i.e., no ad is actually an improvement), then our probability of errantly having a test pass is

$$\Pr[\text{Some ad passes}] = 1 - \Pr[\text{Every ad fails}]$$
$$= 1 - \left(1 - \frac{\alpha}{n}\right)^n$$
$$= 1 - \left(1 - n\frac{\alpha}{n} + \{\text{higher order terms}\}\right)$$
$$\approx \alpha$$

The Bonferroni correction is the most common multiple hypothesis correction that you will see, but you should be aware that it is unnaturally stringent. In particular, it assumes that every test you run is independent of every other test you run: if test X does not pass, then test Y is still as likely to pass as it ever was. But that often isn't the case, especially if your different tests are related in some way. For example, say that you have a database of people's height, weight, and income, and you are trying to see whether there is a statistically significant correlation between their income and some other feature. Height and weight are strongly correlated between people, so if the tall people aren't richer, then the heavy people won't be either: they are more or less the same people! Intuitively, you want to say that we are testing 1.3 hypotheses or something similar, rather than two. Bonferroni would apply if you tested height on one sample of people and weight on an independent set. There are other, complicated ways to adjust for situations such as this, but data scientists are unlikely to need anything beyond Bonferroni corrections.

19.5 Parameter Estimation

Hypothesis testing is, as I mentioned earlier, about measuring *whether* effects are present, but not about quantifying their magnitude. The latter falls under the umbrella of "parameter estimation," where we try to estimate the underlying parameters that characterize distributions.

In parameter estimation, we assume that the data follows some functional form, such as a normal distribution with mean μ and standard deviation α. We then have some method to estimate these parameters, given the data, as follows:

$$\hat{\mu} = \frac{1}{N}\sum_{i=1}^{N} x_i$$

$$\hat{\sigma} = \sqrt{\frac{1}{N}\sum_{i=1}^{N}\left(x_i - \hat{\mu}\right)^2}$$

In this case, these numbers are called the sample mean and the sample variance.

To step back and cast this in statistical terminology, $\hat{\mu}$ and $\hat{\sigma}$ are both test statistics that we calculated from the data. If we threw out our dataset and gathered a new one of the same size from the real-world distribution, $\hat{\mu}$ and $\hat{\sigma}$ would be a little bit different. They have their own probability distributions and are themselves random variables. The question then is how well we can take these random variables as indicators of the actual μ and σ.

We will talk about the more general problem of confidence intervals later, but for now, there are two pieces of terminology you should know whenever you are talking about estimators:

- Consistency: An estimator $\hat{\mu}$ is "consistent" if, in the limit of having many different points in your dataset, it converges to the real μ.
- Bias: As estimator is "unbiased" if the expectation value of $\hat{\mu}$ is the real α.

Most estimators you might cook up on your own are probably going to be consistent. The $\hat{\mu}$ and $\hat{\sigma}$ are both consistent. However, only $\hat{\mu}$ is unbiased; $\hat{\sigma}$ is on average an underestimate of the true standard deviation.

This surprised me the first time I learned it, so let me explain. It's easiest if you imagine that there are only two data points that we use to find the mean and standard deviation. Then $\hat{\mu}$ will be located exactly in-between them, at the place that minimizes $\sum_{i=1}^{N}\left(x_i - \hat{\mu}\right)^2$. If we had set $\hat{\mu}$ to be anywhere else, then $\sum_{i=1}^{N}\left(x_i - \hat{\mu}\right)^2$ would have been somewhat larger. However, the real μ *is* somewhere else, because our $\hat{\mu}$ is only an estimate of μ. It is an unbiased estimate, so μ is equally likely to be more or less than $\hat{\mu}$, but in either case $\sum_{i=1}^{N}\left(x_i - \hat{\mu}\right)^2$ would get larger. Basically, we are making $\hat{\mu}$ an overly good fit for our data, so on average, the sample deviation will be less than the real-world deviation.

An unbiased estimator of the standard deviation is

$$\hat{\sigma} = \sqrt{\frac{1}{N-1}\sum_{i=1}^{N}\left(x_i - \hat{\mu}\right)^2}$$

This will always be somewhat larger than our original expression, but they converge to the same number as N goes to infinity.

19.6 Hypothesis Testing: t-Test

The *t*-test is a more complicated version of hypothesis testing. It is useful for situations where the thing you measure is a continuous number, rather than a binary coin flip, and you want to assess whether the real-world averages of two distributions are the same. For example, you might have cholesterol measurements for patients who did and did not take a cholesterol-lowering medication. Can we say confidently that the medicine works?

Intuitively, this is a simple problem: draw histograms of each distribution and look at them. If their bell curves are nice and distinct, then the distributions are clearly different. If the bell curves overlap a lot, then either the means are very close

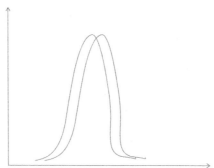

or the spreads are very wide

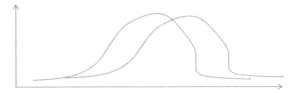

Do this with enough data points and you get to the point where you can easily conclude that there is a difference.

Reducing this intuition to a hypothesis test where we can rigorously calculate a p-value is much, much trickier. It is also much more complicated than the coin-flipping p-value I discussed earlier. So before I give you an idea of how you approach the problem mathematically, I'll show you how to compute it in python: it's very easy. You have two datasets, and the hypothesis that you are testing is that their means are equal. You do *not* know the standard deviations. The math and computation is a lot easier if you assume that the standard deviations are equal, and that assumption is often made by default, but you can also allow them to be different. SciPy's t-test method is called ttest_ind. It produces two numbers: the t-score (which I will explain in a minute) and the p-value we are seeking:

```
>>> from scipy.stats import ttest_ind
>>> t_score, p_value = ttest_ind([1,2,3,4],
[2,2.2,3,5])
>>> t_score, p_value = ttest_ind(
        [1,2,3,4], [2,2.2,3,5], equal_var=False)
```

Now I will give you a brief overview of what's going on under the hood here.

Recall the hypothesis testing process from earlier: formulate the null hypothesis, choose a test statistic that captures the strength of a pattern we have found, and then calculate how much of an outlier that test statistic is. If the test statistic is extremely high, then the null hypothesis is probably bogus.

The t-test gives us two choices of a null hypothesis:

- The two datasets come from the same normal distribution.
- The two datasets from normal distributions with the same means, although the standard deviations could be different.

An important point is that neither of these hypotheses actually gives us a probability distribution. This is in contrast to coin flips, where assuming a fair coin tells you everything there is to know. In a t-test, there are certain things we might want to know that we cannot calculate, because they depend on the parameters that we have not specified.

When formulating the test statistic, the intuition to follow is this:

- Calculate the means $\hat{\mu}_A$ and $\hat{\mu}_B$ of the two datasets and look at how far apart they are.
- If there are more points in our datasets, we expect $\hat{\mu}_A$ and $\hat{\mu}_B$ to be closer to the true μ_A and μ_B. In that case, the difference between $\hat{\mu}_A$ and $\hat{\mu}_B$ should be small if the null hypothesis is true.
- If the datasets themselves have a lot of variance, $\hat{\mu}_A$ and $\hat{\mu}_B$ will be worse approximations to the true μ_A and μ_B. So, the difference between $\hat{\mu}_A$ and $\hat{\mu}_B$ can be larger.

The following test statistic (called the t-statistic) captures the following intuition:

$$T = \frac{\hat{\mu}_A - \hat{\mu}_B}{\sqrt{\dfrac{s_A^2}{N_A} + \dfrac{s_B^2}{N_B}}}$$

where s_A^2 is the sample variance of A and similarly for s_B^2. If you run the math, you will find that the distribution of T will be the same no matter the unknown mean of the underlying distribution and no matter the unknown variances. There is a similar, simpler test statistic that we can use if the population variances are assumed to be equal be unknown.

If the null hypothesis is true, then T will follow what's called a T-distribution. A T-distribution looks very much as a normal distribution with mean 0 and standard deviation 1, but it has somewhat heavier tails. If the t-statistic for our data is unusually large or small, then the null hypothesis is probably not correct.

The t-test is used to test whether the means of two distributions are the same, but it assumes that the underlying probability distributions are normally distributed. You could formulate an equivalent test assuming different distributions if need be, although I've personally never seen it done. A failure of the t-test doesn't necessarily tell you anything about the means; it could also be that your basic assumptions are flawed.

If you want to see whether your data is indeed normally distributed, there is (surprise surprise) a hypothesis test for that: several of them, in fact. One of them based on a metric called the z-score can be used as follows:

```
>>> from scipy.stats import normaltest
>>> z_score, p_value = normaltest(my_array)
```

The z score works by looking at how heavy the tails of the data are and comparing them to the standard deviation of the data around its mean.

19.7 Confidence Intervals

When we are trying to estimate a parameter of a real-world distribution from data, it is often important to give confidence intervals, rather than just a single best-fit number. As I did with the t-test, I will show how to use some pre-canned libraries, the most common use case. I will then back up and discuss what we just did in a more abstract, general way that you can apply to novel problems.

The most common use for confidence intervals is in calculating the mean of the underlying distribution. If you want to calculate something else, you can often massage your data to turn it into a calculation of the mean. For example, let's say you want to estimate the standard deviation of a random variable X. Well, in that case, we can see that

$$\sigma_X = \sqrt{E\left[\left(X - E[X]\right)^2\right]}$$
$$\sigma_X^2 = E\left[\left(X - E[X]\right)^2\right]$$

so we can estimate the standard deviation of X by finding the mean of $\left(X - E[X]\right)^2$ and taking the square root. If we can put confidence intervals on taking a mean, we can put confidence intervals on a lot of other things we might want to estimate.

The typical metric to use is the "standard error of the mean" or SEM. If you see a mean reported as 4.1 ± 0.2, then generally 4.1 is the best-fit mean that you get by averaging all of the numbers, and 0.2 is the SEM. You can calculate the SEM in Python as follows:

```
>>> from scipy.stats import sem
>>> std_err = sem(my_array)
```

If you assume that the underlying distribution is normal, then the SEM will be the standard deviation of the estimate of the mean, and it has a lot of nice mathematical properties. The most notable is that if $\hat{\mu}$ is your sample mean, then the interval

$$\left[\hat{\mu} - z * \text{SEM}, \ \hat{\mu} + z * \text{SEM}\right]$$

will contain the mean 95% of the time, 99% of the time, or any other confidence threshold you want, depending on how you set the coefficient z. Increase that coefficient and you can increase your confidence. The following table shows several typical confidences that people want and the corresponding coefficients:

Confidence (%)	Coefficient
99	2.58
95	1.96
90	1.64

It is very tempting to say that there is a "95% chance that the mean is within our interval," but that's a little dicey. The real-world mean is within your interval or it isn't – you just happen not to know which one. The more rigorous interpretation is to say that if the distribution is normally distributed (no matter its mean and standard deviation), and you take a sample of N points and calculate the confidence interval from them, then there is a 95% chance that the interval will contain the actual mean. The randomness is in the data you sample, not in the fixed parameters of the underlying distribution.

Of course, all of this hinges on the assumption that the real-world distributions are normal. All of the theorems about how to interpret the confidence intervals go out the window when you let go of that assumption, but in practice, things still often work out fine. Statisticians have worked out alternative formulas for the intervals that give the same sorts of guarantees for other types of distributions, but you don't see them often in data science.

The SEM confidence interval is based on the idea that we want to make sure that the true value of the parameter is contained in the interval some known percentage of the time. Another paradigm that you see is that we want the interval to contain all "plausible" values for the parameter. Here "plausible" is defined in terms of hypothesis testing: if you hypothesize that your parameter has some value and calculate an appropriate test statistic from the data, do you get a large p-value?

19.8   Bayesian Statistics

Bayesian statistics, similar to classical statistics, assumes that some aspect of the world follows a statistical model with some parameters: similar to a coin that is heads with probability p. Classical statistics picks the value of the parameter that best fits the data (and maybe adds in a confidence interval); there is no room for human input. In Bayesian statistics though, we start off with a probability distribution over all possible values of the parameter that represents our confidence that each value is the "right" one. We then update our confidence as every new data point becomes available. In a sense, Bayesian statistics is the science of how to refine our guesses in light of data.

The mathematical basis for Bayesian statistics is Bayes' theorem, which looks innocuous enough. For any random variables X and T (which may be vectors, binary variables, or even something very complex), it says that

$$P(T|D) = \frac{P(T)P(D|T)}{P(D)}$$

As stated, Bayes' theorem is just a simple probability theorem that is true of any random variables T and D. However, it becomes extremely powerful when we have T be a real-world parameter that we are trying to guess from the data. In this case, T isn't actually random – it is fixed but unknown, and the "probability distribution" of T is a measure of our own confidence, defined over all the values T, could possibly take.

The left-hand side of Bayes' theorem gives us what we were unable to have in classical statistics: the "probability" that the real-world parameter is equal to a certain value, given the data we have gathered. On the right-hand side, we see that it requires the following:

- $P(T)$: Our confidence about the parameter before we got the data
- $P(D|T)$: How likely the data we saw was assuming that the parameter had a particular value. This often is known or can be modeled well.
- $P(D)$: The probability of seeing the data we saw, averaged over all possible values the parameter could have had.

To take a concrete example, say that we have a user who may be a male or female, but we don't know which. So, we set our prior guess to say there is a 50% confidence they are male and 50% that they are female. Then say, we learn that they have long hair, so we update our confidence to

$$P(\text{Female}|\text{LongHair}) = \frac{P(\text{Female})P(\text{LongHair}|\text{Female})}{P(\text{LongHair})}$$

$$= \frac{\frac{1}{2}P(\text{LongHair}|\text{Female})}{\frac{1}{2}P(\text{LongHair}|\text{Female}) + \frac{1}{2}P(\text{LongHair}|\text{Female})}$$

$$= \frac{P(\text{LongHair}|\text{Female})}{P(\text{LongHair}|\text{Female}) + P(\text{LongHair}|\text{Female})}$$

Personally, I sometimes have trouble remembering Bayes' theorem: the formula for the updated probability is just a little clunky. For me, it's much easier to think about the *odds*, that is, the *relative* probability of female or male. To update the odds, you just multiply by the relative probability of long hair:

$$\frac{P(\text{Female}|\text{LongHair})}{P(\text{Male}|\text{LongHair})} = \frac{P(\text{Female})}{P(\text{Male})} * \frac{P(\text{LongHair}|\text{Female})}{P(\text{LongHair}|\text{Male})}$$

19.9   Naive Bayesian Statistics

The trickiest part of Bayesian statistics is usually calculating $P(D|T)$. This is chiefly because D is usually a multipart random variable in itself – typically, a d-dimensional vector – and the different numbers in it can have very complicated dependency structures. For example, if we want to know whether somebody has diabetes, we might measure their blood glucose and sugar levels, and those two numbers have intricate dependencies that require a deep understanding of biology to model. Or, alternatively, if your model has to learn the relationships automatically, it requires a monumental amount of training data, since all co-occurrences have to be seen.

For this reason, you will often see people use a "naive Bayes" approach. In naive Bayes, we simply assume that all the different variables are independent of each other, conditioned on T. That is,

$$P(D|T) = P(D_1|T)^* P(D_2|T)^* \cdots {}^* P(D_d|T)$$

Now calculating $P(D|T)$ just requires that we have observed enough data to describe each variable's relationship with T in isolation.

The naive Bayes assumption is one of the most dramatic, cavalier oversimplifications in the data science world, but in practice, it's surprisingly effective! The main thing that it tends to do is that if several of the D_i are closely related, they will collectively nudge your classifications too far in one direction. For example, let's say that you're trying to determine whether somebody is a man or a woman. I might then tell you that they have long hair, that they use a hair tie, and that their hair takes a long time to dry. Each of these facts weighs moderately in favor of the person being a woman, so a naive Bayes classifier would become very confident. But really, these facts are just alternative ways of saying that the person has long hair, and there are still plenty of men with long hair (including the author, once upon a time). So, the classifier will probably be right, but it will be overconfident.

19.10   Bayesian Networks

If there are many features in your data, then it is a fool's errand to fully fit a model for $P(D|T)$, with all of the possible interdependencies between the variables. However, there is a happy medium between this impossible task and the dramatic oversimplification of naive Bayes. It is called Bayesian networks.

In a Bayesian network, we allow for some of the D_i to be dependent on each other, and we arrange these dependencies in a graph as the following:

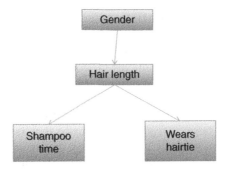

This graph indicates that Gender has a relationship with Hair Length, which perhaps allows us to predict hair length with some accuracy. In many cases, it is a causal relationship, but not always. Similarly, hair length can inform us about how long it takes somebody to use shampoo and whether they wear a hair tie. But that is *it*. Gender might correlate with shampoo time, but only because it correlates with hair length: if you condition on the length of somebody's hair, then gender is independent of shampoo time. Similarly, you can maybe guess whether somebody wears a hair tie based on how long they take to shampoo their hair, but the predictive power goes away if you know their hair length.

Bayesian networks can be very efficient to train and use. In this one, for example, we only need to know the distribution over the possible genders, the distribution on hair length, given the gender, and the distributions of shampoo time and hair tie wearing, given the hair length.

When using a Bayesian network for real applications, you can leverage domain expertise by choosing how to structure the network. Which variables are likely to influence each other, or be independent, or be conditionally independent? Generally, the topology of a Bayesian network is constructed by hand, and afterward, it is trained and evaluated on real-world data.

In a later chapter, we will discuss some of the tools available for training and using Bayesian networks in Python.

19.11   Choosing Priors: Maximum Entropy or Domain Knowledge

If you are trying to train a Bayes classifier, it is easy to extract baseline priors from the training data. However, there are other situations where our ignorance really is complete, so we don't want to feed any preconceived fairy tales into our priors. The instinctive way to handle this in the case of something such as determining somebody's gender is to set the baselines equal: 50% chance woman, 50% chance man.

There is a mathematical theory that justifies this approach: using the prior distribution with the maximum entropy. Previously, we saw that the entropy of a probability distribution measures how much uncertainty there is in it, so choosing a prior that maximizes the entropy reflects complete ignorance.

Recall that if T is a discrete variable with n possible states, then entropy is defined as

$$H[T] = -\sum_{t=1}^{n} p_t \ln(p_t)$$

In this case, $-\ln(p_t)$ measures how "surprising" it is to get a particular result t, and $-p_t \ln(p_t)$ is t's contribution to the overall entropy. If p_t is 1, then $\ln(p_t)$ will be 0, so t would contribute nothing to the entropy. Taking the opposite extreme, if p_t is very small, then the surprise is large, but it happens so rarely that t contributes little entropy. Intuitively, there should be a "sweet spot" that maximizes the entropy, and it turns out that we can get this by setting every p_t to the constant $1/n$.

This definition of entropy has an analog for continuous probability distributions:

$$H[T] = -\int f(t)\ln(f(t))dt$$

Similarly to the discrete case, the maximum entropy distribution is only coherent if there is a finite interval (or set of them) over which it is defined, and it is just a constant value of

$$f(x) = \frac{1}{x_{max} - x_{min}}$$

19.12 Further Reading

Janert, P, *Data Analysis with Open Source Tools*, 2010, O'Reilly Media, Newton, MA.

Diez, D, Barr, C & Çetinkaya-Rundel, M, *OpenIntro Statistics*, 3rd edn, 2015, OpenIntro Inc.

19.13 Glossary

Bayesian network A dependency graph between several random variables, used to model which ones are likely to be conditionally dependent on which others. A good Bayesian network is more powerful than Naive Bayes, but still sparse enough that it can be effectively trained on data.

Bonferroni correction A way to adjust the required *p*-value for testing a hypothesis if you are testing multiple hypotheses at the same time. It accounts for the fact that, while each individual hypothesis is unlikely to pass by pure chance, the probability of *some* hypothesis passing by dumb luck increases as you test more hypotheses.

Consistent estimator An estimator that, in the limit of having a lot of data, is guaranteed to converge to the correct value (assuming that the real-world distribution is, indeed, of the same family that we are assuming).

Entropy A measure of how hard it is to predict the outcome of a probability distribution. Often, we pick priors to maximize the entropy in Bayesian statistics.

Estimator A test statistic that is used to estimate a parameter in the probability distribution that data is assumed to have been drawn from.

Hypothesis testing A framework for testing whether a pattern in the data is "statistically significant," by looking at how likely it is to occur by random chance in the absence of an underlying real-world phenomenon.

Multiple hypothesis testing Hypothesis testing when there are multiple hypotheses that could be correct, such as multiple medicines all of which are being tested for clinical effectiveness.

Naive Bayes The assumption that all features in a dataset are independent of each other when you condition on the target variable. This makes training Bayesian models vastly simpler, and they are often still surprisingly effective.

Null hypothesis In hypothesis testing, the null hypothesis is the assumption that whatever pattern we have found in the data in just a fluke. For example, let's say that we have 9 of 10 flips of a coin have been heads and we think that the coin might be biased. The null hypothesis holds that the coin is fair.

p-Value The probability of seeing a result that is as extreme (or more so) as what we see in the data if the null hypothesis is true.

Prior In Bayesian statistics, this is our confidence distribution over an unknown variable before we have acquired any data.

t-Test A hypothesis test for determining whether the means of two distributions are different.

Test statistic Any number that is calculated from a dataset.

Unbiased estimator An estimator that is on average equal to the correct underlying value. The same mean from a dataset is an unbiased estimator of the real mean. However, the sample variance is *not* an unbiased estimator of the true variance: it is systematically biased to be smaller.

20

Programming Language Concepts

So far, the book has focused on quick-and-dirty scripting, in the service of larger analytics goals. The most in-depth I've gotten about code is how to do unit testing and work within the context of a conventional software engineering team.

This chapter will take a step back and get into some of the more abstract, theoretical aspects of programming languages. This is important to know for two reasons. First, these considerations will often dictate important decisions about what tools to use when. You don't want to lock yourself into using the wrong technology for a task, especially if you find yourself working to create a large software framework. Secondly, tools that are fundamentally different can take some getting used to; understanding the core concepts will ease the transition if you have to pick up a new tool that is profoundly different from what you're used to.

20.1 Programming Paradigms

A "programming paradigm" is a conceptual way to think about the logical structure of a program and implement it in code. You can think of it as a sequence of instructions for how to perform the computation, a mathematical specification of what the output should look like, or a range of other options.

Before I get into the details, the first big thing you should know is that most modern high-level languages support all of these paradigms to one degree or another. This means that to a large degree you can mix and match paradigms depending on what works best for the problem at hand.

Strictly speaking, all of these paradigms are equivalent; any computation that can be done with one can be done with any of the others. Conceptually though, they can be quite different to think about, lending themselves to different applications and even personal temperaments.

Some people are pretty dogmatic about which paradigms are best, and some language shoehorns you into one in particular. A number of them have fancy

The Data Science Handbook, First Edition. Field Cady.
© 2017 John Wiley & Sons, Inc. Published 2017 by John Wiley & Sons, Inc.

theoretical aspects that can be very useful in certain situations. They also tend to be different in terms of the performance and maintenance of the code.

The big three paradigms that you will see are often called "imperative," "object-oriented," and "functional," and I will introduce them all in this chapter. Python has at least partial support for all of them.

20.1.1 Imperative

In imperative programming, your code is mostly a sequence of instructions for the computer to follow. These could be things such as appending a new element to a list, overwriting an existing variable in memory, or writing a file.

The following Python code is an imperative way to read in a CSV file containing demographic information and then calculate the average age of all people in each state:

```
lines = open('data.txt')
broken_lines = [l.split(',') for l in lines]
ages_by_state = {}
for bl in broken_lines:
 state, age = bl[2], bl[5]
 state = state.strip().lower().replace('.','')
 age = float(age)
 if state in ages_by_state.keys():
  ages_by_state[state].append(age)
 else: ages_by_state[state] = [age]

mean_by_state = {}
for state, ages in ages_by_state.items():
 avg_age = sum(ages) / len(ages)
 mean_by_state[state] = avg_age

out_lines = [state + ',' + str(age)
    for state, age in state_age_pairs]
output_text = '\n'.join(out_lines)

open('output.txt','w').write(output_text)
```

20.1.2 Functional

Functional programming is largely inspired by the desire to avoid "side effects." A side effect means any modification that is done to existing variables (such as appending an element to a list or incrementing a number) or any interaction of the program with the outside world (such as printing to the screen). So, the following would be side effects:

```
print 'Hello world!'
a = a + 1
```

Obviously, we ultimately want our code to actually do something, so side effects are a good thing in general. The problem though is that they make it difficult to reason about our code. The order in which steps are taken matters (did we print the number before or after we incremented it?), and it can get very hard to keep track of sometimes.

I've run into this personally when using the Python interpreter to help me fiddle with scripts I was writing. I had a collection of variables that I was trying to massage so that they would fit into a statistical model I was building. I would run and rerun portions of my script, making small changes in between runs as I tweaked the logic. Eventually, my variables were in good shape, so I thought the code was working correctly. But when I reran the code from scratch, things broke again. It turned out that some of the necessary changes had been made by previous versions of the code, and I had forgotten that those side effects had already happened.

In functional programming, your code is broken up into "pure" functions, which take some input (or maybe none at all) and return an output, but have no side effects. Here is a much more functional version of the same code from the previous section:

```python
def normalize_state(s):
  return s.strip().lower().replace('.','')

def mean(nums):
return sum(nums) / len(nums)

def extract_state_age(l):
  pieces = l.split(',')
  return normalize_state(state), float(age)

def get_state_mean(pairs):
  ages_by_state = {}
  for age, state in pairs:
    ages_by_state[state] = \
      ages_by_state.setdefault(state,[]) + [age]
  state_mean_pairs = [(state, mean(ages))
      for state, ages in ages_by_state.items()]
  return sorted(state_mean_pairs, key=lambda p: p[1])

def format_output(state_age_pairs):
  out_lines = [state + ',' + str(age)
    for state, age in state_age_pairs]
  return '\n'.join(out_lines)

lines = open('data.txt')
state_age_pairs = [extract_state_age(l) for l in lines]
output = format_output(state_age_pairs)
open('output.txt','w').write(output)
```

In this version, all but the last four lines of code are just defining functions. In the final four lines, I do create some variables based on the already existing ones, but I don't modify any that were already created. The only side effect is in the last line. Technically, within the get_state_mean function, I modify the variable ages_by_state, so this isn't 100% functional code at every level, but you get the idea.

The real nirvana of functional coding though isn't just arranging your code into pure functions. It's treating functions as variables in their own right, which can be passed as arguments into other functions or even generated on the fly. Take this code for example. It is a function that takes in a date encoded as a string, figures out the way the string is formatted, and returns a function that will extract the year out of strings with the same format:

```
def get_year_extractor(example_date):
    # not sure if example_date is YYYYMMDD
    # or YYYY-MM-DD or MM-DD-YYYY
    if len(example_date)==8: return lambda d: d[:4]
    elif (example_date[4]=='-'
        and example_date[7]=='-'):
        return lambda d: d[:4]
    else: return lambda d: [-4:]
extract_year = get_year_extractor(dates[0])
years = [extract_year(dt) for dt in dates]
```

Or, try the following, which is more along the lines of what you see in analytics:

```
def get_stats(data_lst, f):
    def valid_input(x):
        try:
            _ = f(x)
            return True
        except: return False
    vals = [f(d) for d in data_lst if valid_input(d)]
    return {'num_valid': len(vals),
        'num_invalid': len(data_list)-len(vals),
        'mean': sum(vals) / len(vals),
        'max': max(vals)
}
```

In this case, you might experiment with a wide range of custom-built functions, plugging each one into get_stats to see how they all perform.

Functional programming lends itself well to data science, especially if you are doing some kind of batch processing. There is a clear sequence of logical stages leading from your raw data up through your final analytics outputs – no user interactions or other things that are hard to think about as pure functions. It is easy to plug-and-play with different functions, passing them around to operate on your data in novel ways. In the Big Data space, Spark is a functional framework. This is part of why many people (myself included) try to write code in as functional a way as possible.

An oft-touted advantage of functional programming is its performance, at least if the code is compiled rather than interpreted. A functional program is more of a mathematical specification for a program's logic than a set of instructions for the computer. In theory at least, the compiler can look at those specifications and figure out a highly efficient way to implement them – much better than you probably would have if you'd written your code in an imperative way.

In practice though, the compiler isn't usually very smart, and it often implements things in a fairly naive way. If your functional code is in an interpreted language, it will also certainly be naively implemented and pretty slow.

There are also certain control structures where functional syntax can become a bit clunky (or at least unintuitive the first time you see it). Check out these two pieces of code, which do the same thing imperatively and functionally:

```
# Imperative version
my_variable = initial_version
while not my_stopping_condition(my_variable):
    my_variable = my_function(my_variable)
```

```
# Functional version
def loop_as_function(variable):
    if my_stopping_condition(variable): return variable
    else: return loop_as_variable(my_function(variable))
my_variable = loop_as_function(initial_version)
```

Doing 100% functional programming requires that you replace loops with recursion in this way, which many people regard as an eyesore. It's also grotesquely inefficient to do in Python because of the under-the-hood boilerplate required for all the nested function calls (it can be much faster in compiled functional languages such as Haskell, which use something called the "tail call optimization," but Python doesn't have that feature).

The granddaddy functional programming language is Lisp, which dates back to 1958 but still has devoted adherents. Popular, mostly functional languages today include Haskell and ML. Scala famously emphasizes its functional programming support, although it can also be written in a very object-oriented way.

20.1.3 Object-Oriented

An object-oriented language will package data and the logic that handles the data into user-friendly black boxes called "objects." When using an object, you don't need to worry about how the data is structured or how to untangle that structure; you only interact with special-purpose functions called "methods" that the object presents to you.

Python is inherently an object-oriented language. Everything you ever use or define in Python code, including variables, functions, and even libraries that you import, is an object. Every action you ever take in Python is calling a method on some object.

You might object that simple things such as integer addition are not object methods. Actually though, "+" is just syntactic sugar around the "__add__" method, as you can see here:

```
> x, y = 4, 5
> x + y
9
> x.__add__(y)
9
```

The interesting thing about Python is that, even though it's technically 100% object-oriented, you don't have to write your code that way. If you read my code, you'll see that I very rarely define my own classes; I do almost everything with Python's built-in container objects (such as lists and dictionaries) and the objects supplied by the libraries I use. The fact that everything gets implemented as an object is incidental; my code reads like a blend of imperative and functional. This is partly a matter of personal taste and partly a reflection of the kind of work I usually do.

Where object-oriented coding shines is having your code interact with external resources. Something such as a GUI, where the user presses buttons at will, is best thought of as an object. The object carries around all of the data you're working with, the specification of how the GUI is laid out, and whatever boilerplate is required to interact with the computer's graphics. Every button press calls some method on the object, which changes the object's internal state (i.e., a side effect) and takes any other actions necessary. If you have two processes that interact with each other, if your code connects to an iPhone or an interactive website, or if the user gets brought into the loop – all of these situations are best thought of as interactions between objects, via their methods.

Everything in Python is an object, but generally, they're a prepackaged type of object such as a list or an int. The key feature of object-oriented code is the definition of whole new types of objects. These types of objects are called "classes." A class specifies the internal structure of an object, as well as all of its associated methods. You can have many objects that are all of the same class, but the class itself is only defined once. To see how this works, let's dive into some code:

```
class Person:
    def __init__(self, name, age):
```

```
    self.name = name
    self.age = age
  def talk(self):
    print 'My name is %s and I am %s years old' \
        % (self.name, self.age)
Janice = Person('Janice Smith', 28)
Bob = Person('Bob Jones', 30)
Bob.talk()
```

This code defines a class called "Person," with methods "__init__" and "talk," and creates two Persons called Janice and Bob.

The first thing that might jump out at you is the goofy word "self." Where did that come from? It looks like it's an argument in the functions __init__ and talk, but it's never there when those objects get called. What gives? Ok, this aspect of Python is incredibly confusing for many people, so much so that some of them abandon the language entirely, so let me try to make it clear.

When you, the programmer, call a member function on an object, you call it my_object.my_function(argument1, argument2). However, under the hood, the actual object itself gets passed into the function, silently, as an additional argument before all of the others. So, when you write the code that actually implements the function, you write it so as to accept this additional argument. Calling the argument "self" is just a convention, but it's an almost universal one.

If you want to refer to your Person's age within the member function, you say "self.age." In many other object-oriented languages, you just say "age," and it is understood that this refers to the "age" field of the object being operated on (these language also don't require the object to be an argument in the implementation of the function). However, I think that Python's way, while admittedly more verbose, makes the code much more readable. In a long and complicated file, it's entirely possible that I declared a variable called "age" somewhere else in my script or imported a library called "age." Especially if somebody else implemented the Person object and I'm just perusing their code, I might not know that Person even has "age" field. So, I may end up having to dig through the rest of the code just to figure out whether "age" is a member of the object or whether it's something else. Saying "self.name" removes all of this ambiguity. You are welcome to hate this aspect of Python if you so choose; many great engineers will agree with you. But personally, I love it.

Now that you understand what "self" means, the "talk" method should look pretty straightforward. However, __init__ might be something of a mystery. This special-purpose function is what builds an object up when it is first created. When I said "Person('Janice Smith', 28)," the first thing that happened was an empty Person object was created. Then, under the hood, its __init__ function was called with itself, the string "Janice Smith," and the number 28 as its arguments. The __init__ function then created the new "name" and "age" fields for this object and populated them accordingly.

You might hear __init__ referred to as a "constructor," which serves the same purpose in other languages. Technically, this is incorrect though. A constructor aids in actually constructing the object when it first comes into being. In Python though, the object technically exists before __init__ is called; it just hasn't been filled with any data fields yet.

The place where objects and classes start to get really interesting is when "inheritance" comes into play. I might define an "Animal" class, with methods that apply to all animals. But then I might also want a "Parrot" class, which carries over all of the logic and member functions of Animal, but adds on others that are specific to parrots. Any Parrot would then be both a Parrot and an Animal. In Python, the code might look like this:

```
class Animal:
 def __init__(self, name):
  self.name = name
class Parrot(Animal):
 def talk(self):
  print self.name + ' want a cracker!'
Fido = Animal('Fido')
Polly = Bird('Polly')
```

In this case, the internal structure of a Parrot is the same as that of a generic Animal; the only difference is that Parrot has a "talk" function. If I wanted to add additional internal structure to Parrot, I could do it the following way:

```
class Animal:
 def __init__(self, name):
  self.name = name
class Parrot(Animal):
 def __init__(self, name, wingspan):
  self.wingspan = wingspan
  Animal.__init__(self, name)
 def talk(self):
  print self.name + ' want a cracker!'
Fido = Animal('Fido')
Polly = Bird('Polly', 2)
```

In this case, we have given Parrot its own __init__ function, which takes in an additional argument and has precedence over the __init__ function of the Animal class. In some object-oriented languages, both __init__ functions would then get called when we create Polly. In Python though, Parrot's __init__ function overrides Animal's entirely, and it is up to the coder to make sure that everything they really wanted from Animal.__init__ is salvaged. In this case,

Animal.__init__ doesn't do anything that conflicts with Parrot.__init__, so we are free to call Animal.__init__ explicitly within Parrot.__init__. As with the "self" word in member functions, this makes your code more verbose and arguably uglier, but it also makes the logic much more explicit.

Personally, my code generally doesn't use class inheritance. The one big exception is that some of Python's libraries provide a very fancy class that wraps a lot of functionality, and you write your own class that inherits from it. For example, the HTMLParser class has logic for wading through and parsing HTML text. I've often written classes that inherit from HTMLParser and that identify and process specific pieces of information as HTMLParser walks through the text.

Similar to functional programming, the object-oriented paradigm has a broad theoretical foundation and many people who are real purists about using it. Hardcore object-oriented code consists almost entirely of defining class and their member functions and makes liberal use of class inheritance and other fancy features. A few other things you should be familiar with:

- Static member data are not associated with individual objects, but with the class itself. For example, there might be a static integer that keeps track of how many instances of a class are currently in existence.
- It is possible to have a single class inheriting from multiple other classes. For example, a Parrot can be both an Animal and a ThingThatFlies. You can do this in Python and many other languages, but it's relatively rare.

20.2 Compilation and Interpretation

"Compiled languages" and "interpreted languages" used to have very clear-cut meanings. Advances in recent years though have blurred the distinction into more of a continuum. First, I'll go over the old-school meanings and then get into some of the modern variants.

Traditionally, a compiler is a special-purpose program that translates human-readable code into "machine code," which just operates on raw bytes and is directly executed by the microprocessor. In performing the translation, the compiler has to know the details of the microprocessor it's writing code for and how it expects its machine code to be formatted. The compiler must make all sorts of judgment calls about how a single high-level operation should be translated into a series of low-level operations. Some compilers are pretty naive in the way they make these judgment calls, but so-called optimizing compilers make them in very judicious ways and can even reorganize much of your program so that the behavior is the same, but it runs more efficiently.

A language such as C is compiled, so your computer can't "run" your C code in any direct fashion. First, you have to compile your code into machine code.

The resulting blob of code is called an "executable," since the computer can actually run it. However, the executable contains no vestige of the fact that it was originally written in C. The original machine code could have been written in any compiled language, or even directly in machine code one byte at a time, and have the executable be the same.

An important concept to understand with compiled languages is the difference between something happening at "run time" and at "compile time." A syntax error in your code will generally show up as soon as you try to compile it. The compiler will be unable to perform its translation and will throw an error, and you find out sooner rather than later that you screwed something up. Other times though, the code will compile just fine, and you will only discover the problem when the computer actually tries to run the machine code. This discovery might happen hours into the run or maybe even after you've pushed a product to consumers. For this reason, a lot of work in compiler design boils down to trying to find issues at compile time.

In contrast to compiled languages such as C, you have interpreted languages such as Python. An interpreter is a special-purpose program that reads and executes your code one line at a time. The interpreter itself is a blob of machine code that was originally written in something such as C. Python code is acted on by the interpreter program but never translated into machine code.

Generally speaking, compiled languages are blindingly fast and efficient relative to interpreted ones. This is partly because interpreted language must incur the overhead of parsing every line of code at run time and partly because there is no opportunity for a compiler to build optimizations in. However, interpreted languages let you play around and debug your code one line at a time, and they spare you from worrying about what's going on under the hood.

If somebody wanted, they could write an interpreter for a compiled language such as C and, similarly, a compiler for an interpreted language such as Python. This has happened in some cases for some languages, but it's pretty rare. An interpreter for something such as C would completely kill performance, which is the whole reason to use C in the first place. If you wanted to compile Python, the compiler cannot always infer the key things about the program (such as which variables are which types) that allow for efficient compiled code.

Ok, that's how things used to be. Now the gray area. I actually lied when I told you that Python is interpreted. Many languages in the past were really and truly interpreted, but Python can be translated into an intermediate representation called "bytecode." Python will be run one line at a time if you open up an interpreter from the command line. But if you run your Python script all at once, or if a library is imported, it is first translated into bytecode. This is far from compilation in the traditional sense, since bytecode is at the same level of abstraction as raw Python and gives minimal performance optimizations

An even more gray area is the Java language. Java also compiles to an intermediate bytecode, but Java bytecode is very low-level relative to the original

Java language. In fact, the interpreter is called the "Java virtual machine" (JVM), since the bytecode feels less like a regular programming language and more like low-level machine code. The Java compiler does a ton of optimizations in generating the bytecode, so you get most of the performance benefits of a compiled language. The JVM mostly just provides you with memory management and access to the computer's resources.

The .NET framework that Microsoft uses is very similar to the JVM. The C# language is essentially equivalent to Java (originally, they were planned to be the same language, but then corporate politics got in the way), and it compiles into .NET bytecode in the same way.

For both the JVM and .NET, there are actually a wide range of languages that can be compiled into the same bytecode, even if Java and C# are the flagships. So, bytecode is really becoming the lingua franca of these software environments; the actual machine code is an afterthought that is taken care of by the virtual machine, in a way that is specific to whatever computer it's running on.

The blurring of compilation and interpretation has even taken place on the level of the guts of the computer. Without getting into too much historical detail, Intel changed their fundamental machine code from what's called a CISC model (specifically x86) to a more efficient RISC model. However, doing this rendered all previously compiled code obsolete, so they hard-coded logic into the actual silicon that translated compiled x86 code into the RISC code, in real time as the executable was running.

So really, you should think of "compilation" as just the translation of code that is in one language into some lower-level language. Conversely, you should then think of "interpretation" as plugging code into a "machine" (virtual or physical) that executes the code.

One final complication bears noting. "Just-in-time" (JIT) compilation sometimes happens on the fly, while an otherwise interpreted language is being run. Maybe the compiler couldn't guarantee, at compile time, that a list would only ever contain integers. However, as the program is running that guarantee might become possible, and the code that processes the list can be recompiled on the fly with the new knowledge. This is a lot of overhead, but sometimes, JIT compilation gives huge performance gains.

20.3 Type Systems

Besides the programming paradigm(s) that it implements, a language is also characterized by its "type system." Every variable defined in your code will have a type associated with it, such as an integer, a dictionary, a string, or a custom-defined class. Performing an operation on an object requires knowing what type it is, so that we know how to parse the object's underlying bytes and actually implement the operation. There is a large, theoretical discipline devoted to studying type systems and defining their properties, but you don't need to know about it. There are really just two main concepts you should be familiar with: whether something is "strongly" or "weakly" typed and whether it is "statically" or "dynamically" typed.

20.3.1 Static versus Dynamic Typing

A language is "statically typed" if the computer figures out, at the time the code is compiled, what the type is of all the variables. This allows the compiler to store and process the data in the most efficient way possible. It is dynamically typed if the types are not known until the code is run, meaning that there will be some additional boilerplate to keep track of what variables are integers, strings, lists, and so on.

Python is a great example of a dynamically typed language. The interpreter is written in C, and under the hood, every variable is implemented as a C structure called a PyObject. One function of the PyObject structure is to keep track of what the type is of each variable. There is a lot of overhead in this approach. Most simply, you have to store more stuff in RAM: not just your actual data, but the type metadata. The other problem is that, before your code can perform some operation (such as "+") on a variable, it must first check what data type that variable is and hence what the operation means in this context. Dynamic typing has many benefits in terms of flexibility, but you pay a large performance cost.

In a statically typed language such as C, on the other hand, the compiler can just translate every operation into the appropriate byte-level manipulations, without storing any explicit reference to the data types or any method lookups.

Many interpreted languages have a particular type of dynamic typing sometimes called "duck typing." This means that a variable is considered to be the "right" type if every operation that is ever called on it is defined when it is called. The term "duck typing" comes from the idea that if it has a quack() method and a walk() method, we may as well call it a duck.

In compiled languages, one of the most important steps is for the compiler to figure out what data type every variable is, because that will dictate how every subroutine should operate at the level of byte manipulations. In many languages, such as C, you must do this explicitly and tell the computer what type every variable is (and hence, implicitly, how it should store that variable in bytes). In some more modern compiled languages, you don't have to declare the types explicitly, and there are elaborate mechanisms in the compiler that examine your code and infer what types the variables are.

20.3.2 Strong versus Weak Typing

Typing strength is a much fuzzier notion than whether a language is dynamically or statically typed. Roughly, it means to what degree the language forces you to use types and their operations consistently. Let me give you several examples:

- Combining strings with other types. Python is strongly typed in that

```
>>> c = "hello" + 5
```

  will throw an error. However, many other languages will guess that you mean to turn 5 into a string and set c to "hello5."
- Mixing integers and floats. Python is weakly typed in that "3/2" and "3.0/2" are both valid expressions, but they will return different values (1 and 1.5, respectively) because integer and floating-point division are different operations. Python guesses that in 3.0/2 you actually mean 2.0 and performs the division accordingly. However, a language such as OCaml will actually give a compile-time error; you have to explicitly turn 2 into 2.0 before diving 3.0 by it.
- C is typically a strongly typed language, since you have to declare the type of every variable in the code itself and the code won't compile if the types don't match up. However, you can also force the computer to treat a random chunk of bytes as if they were data of a particular type. The random bytes may or may not constitute a valid instance of that type.
- Remember that the underlying bytes of memory are not typed; C lets you access this flexibility, whereas higher-level languages put layers of security around it.

20.4 Further Reading

1 Scott, M, *Programming Language Pragmatics*, 4th edn, 2015, Morgan Kaufmann, Burlington, MA.
2 Martin, R, *Clean Code: A Handbook of Agile Software Craftsmanship*, 2009, Prentice Hall, Upper Saddle River, NJ.

20.5 Glossary

Anonymous function A function that is never given an explicit name. These are often defined on the fly when they are passed as an argument into another function.

Compiler A software program that translates human-readable source code into a low-level language that is more suitable for actually running. This is often machine code or bytecode for a virtual machine.

Constructor A subroutine that constructs a user-defined object in memory, including correctly initializing all of its internal state.

Duck typing A type system where a variable is considered to have the correct type if every operation that is ever called on it is defined.

Functional programming A programming paradigm where the code consists mostly of pure functions.

Imperative programming A programming paradigm where the code consists mostly of side effects.

Inheritance In object-oriented programming, this is defining a new class that inherits the logic and methods of a previously existing class.

JIT compilation Just-in-time compilation.

Just-in-time compilation Compiling parts of an interpreted language at runtime, so as to gain a performance advantage.

Object-oriented programming A programming paradigm where your code consists mostly of defining new classes and objects, which mask their internal state and present APIs that can be accessed by other objects.

Programming paradigm A way to think about the operation of your program and to break it apart into logical pieces.

Pure function A subroutine that returns a value and has no side effects.

Side effect Any change in a program's state that is caused by a function executing. Side effects include modifying existing variables in memory and writing output to the screen or the file system.

Type safety Enforcing that operations within a piece of code only ever happen to variables of the appropriate type.

Virtual machine A piece of software that interprets low-level bytecode (such as Java bytecode) and handles the interface between it and the hardware and operating system.

21

Performance and Computer Memory

This chapter discusses ways that your code can be made to run faster. It roughly breaks into two very distinct topics:

- The theory of how fast an algorithm is in the abstract, independent of the details of the computer or the implementation. You don't need to know a whole lot about this subject – mostly just how to avoid a few potentially catastrophic errors.
- Various nitty-gritty performance optimizations, which mostly involve making good use of the computer's memory and cache.

The first of these topics relates to figuring out which algorithms are fundamentally, theoretically better than others. The second topic is about how to eke out real-world performance gains for whatever abstract algorithm you are using.

21.1   Example Script

The following script examines two different ways to solve the same problem: given a list of items, count how many of the entries are duplicates of some other entry. It sees how long each algorithm takes for lists of varying length and plots out the time.

The two ways I solve the problem are as follows:

- For each element x in the list, count how many times it occurs in the list – if it occurs more than once, then it is a duplicate. The key thing here is that counting the occurrences of x will require searching through the whole list and checking whether the elements are equal to x. That is, for every element of the list, we loop through the whole list.
- Keep a dictionary that maps elements in the list to how many times they have occurred, then loop through the list, and update this dictionary. At the end, add up all of the counts that are greater than 1.

The Data Science Handbook, First Edition. Field Cady.

For reasons that will become clear, I call these approaches O(n^2) and O(n), respectively.

```python
import time
import matplotlib.pyplot as plt
import numpy as np
def duplicates_On2(lst):
    ct = 0
    for x in lst:
        if lst.count(x) > 1: ct += 1
    return ct

def duplicates_On(lst):
    cts = {}
    for x in lst:
        if cts.has_key(x):
            cts[x] += 1
        else: cts[x] = 1
    counts_above_1 = [ct for x, ct in cts.items()
        if ct > 1]
    return sum(counts_above_1)

def timeit(func, arg):
    start = time.time()
    func(arg)
    stop = time.time()
    return stop - start

times_On, times_On2 = [], []
ns = range(25)
for n in ns:
    lst = list(np.random.uniform(size=n))
    times_On2.append(timeit(duplicates_On2, lst))
    times_On.append(timeit(duplicates_On, lst))

plt.plot(times_On2, "--", label="O(n^2)")
plt.plot(times_On, label="O(n)")
plt.xlabel("Length N of List")
plt.ylabel("Time in Seconds")
plt.title("Time to Count Entries w Duplicates")
plt.legend(loc="upper left")
plt.show()
```

There is noise in the data because I ran it on my actual computer – there are other processes going on that can make this take more or less time. However, some trends are still clear. When the lists are short, the two algorithms are

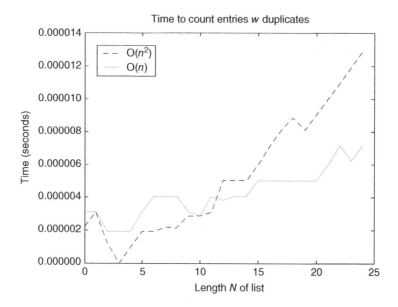

comparable, with $O(n^2)$ often being better. But as the lists get longer, a gap opens up, $O(n^2)$ starts to take much longer. I tried it again, with lists up to length 50, and got the following:

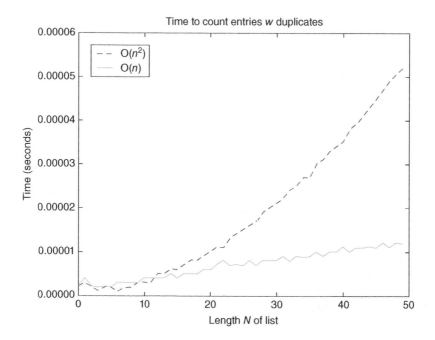

We can see that as the list gets longer, $O(n)$ becomes a better way to solve the problem. This has nothing to do with the computer I'm using or the fact that I'm doing it in Python: it is a fundamentally better algorithm.

21.2 Algorithm Performance and Big-O Notation

The standard way to discuss the theoretical performance of an algorithm in the abstract is called Big-O ("Big oh") notation. In the aforementioned example, we had an algorithm that is $O(n)$ and one that is $O(n^2)$. Intuitively saying that an algorithm is $O(n^2)$ means that, in the limit of large input, the algorithm's runtime will be approximately some constant times n^2. Similarly, if the algorithm is $O(n)$, then it will be approximately a constant times n. We can see this visually in the given graph: one curve is more or less a straight line, and the other resembles a parabola.

There are two key things about assessing algorithms using Big-O notation:

- How long the code actually takes involves a lot of other factors, such as the machine you are running on and how efficiently each step in the algorithm is done. In the aforementioned example, $O(n)$ algorithm actually takes longer for small n, because I used a particularly inefficient implementation.
- In the limit of large input size, differences in Big-O notation will come to dominate the runtime between two algorithms.

Basically, Big-O notation tells you nothing about how long one step in your algorithm will take, but it does tell you how the number of steps grows as a function of input size.

The Big-O performance of an algorithm is often called its "complexity." This is an unfortunate choice of terminology, since the number of steps an algorithm takes has nothing to do with how complicated those steps are. In fact, in my experience, the less "complex" algorithms are often the most complicated to understand, because people do all kinds of complicated tricks to reduce their Big-O complexity. But the terminology is universal, so I will use it.

Look more closely at the $O(n^2)$ algorithm. There were n elements in the list that we looped over. For each element x we called the count() method of the list, which looped over the list checking for equality with x. This means that there are n^2 comparisons made. Since they are made in order, the time for the code to run will be approximately the time required for a comparison times n^2.

The other algorithm also loops through the list. However, for a given element x, it only checks whether x is in the cts dictionary and updates that dictionary as needed. Under the hood, dictionaries are implemented such that this operation takes a fixed amount of time, regardless of how many elements are in it. That is, checking and updating the dictionary takes $O(1)$ time. So, there are n

iterations of the outer loop, each of which takes O(1) time, for a total runtime that is O(*n*).

As a rule of thumb, your data science code should never have an O(n^2) algorithm that operates on your entire dataset, unless you *really* mean it. Every basic data science operation I've shown you so far in the book (training classifiers, different relational operations, etc.) has good asymptotic performance when it's implemented well (and the libraries I'm showing you all have efficient implementations), so chances are that you wouldn't run an O(n^2) algorithm by accident. It happens though, and it is disastrous when it does. If you end up writing core algorithms yourself, you will have to be very keenly aware of their complexity.

21.3 Some Classic Problems: Sorting a List and Binary Search

This section will show you, by example, how to calculate the Big-O complexity of an algorithm by two classic problems: searching for an element in a list and sorting a list.

Let's say you have a list of length *n* and you want to find whether there is an element *x* in it. Then, you might implement the following algorithm:

```
Input:
    List L of length n
    x
Initialization:
    i = 0
Algorithm:
    For i in 1, 2, …, n:
        If L[i] == x: Return True
    Return False
```

We assume that getting L[*i*] and comparing it to *x* is an O(1) operation, and we see that this can happen up to *n* times. So, the search algorithm is O(*n*).

Now let's assume that the list is sorted. This assumption allows for a much more efficient algorithm called a binary search. It goes as follows:

```
Input:
    List L of length n
    x
Initialization:
    i = 0
    j = n-1
```

```
Algorithm:
    While True:
        y = L[(i+j)/2]
        If y== x: Return True
        Elif j==i or j==i+1: Return False
        Else:
            If y>x: j = (i+j) / 2
            Else: i = (i+j) / 2
```

This algorithm's complexity is more difficult to analyze. The key observation is that after each iteration of the main loop, the distance between i and j divides by 2, and then we rerun the loop on this smaller problem. If we let $T(n)$ be the runtime for a given n, we then see that

$$T(n) = O(1) + T\left(\frac{n}{2}\right)$$

We can in turn expand the $T(n/2)$ out to see that

$$T(n) = O(1) + O(1) + T\left(\frac{n}{2^2}\right)$$

$$T(n) = O(1) + O(1) + O(1) + T\left(\frac{n}{2^3}\right)$$

$$T(n) = O(1) + \ldots + O(1) + T\left(\frac{n}{2^k}\right)$$

Every time you expand $T\left(\dfrac{n}{2^k}\right)$ out, you incur an additional $O(1)$ runtime, and you can continue the expansion until $\dfrac{n}{2^k}$ becomes $O(1)$. This will happen after $\log_2(n)$ steps, so we can see that the binary search is an $O(\log(n))$ algorithm. For large n, this will be MUCH faster than $O(n)$.

The efficiency of binary search raises the question of how long it takes to sort a list. The most obvious algorithm is

```
Input:
    List L of length n
Initialization:
    NewList = Empty list
Algorithm:
    While L contains elements:
        mn = min(L)
```

```
    delete mn from L
    NewList.append(mn)
Return NewList
```

Let's assume that calculating mn = min(L) requires looping over all of L. Then, assume that deleting it from L takes O(1) time. In this case, the loop will first calculate the min of a list of length n, then it will be a list of length $n-1$, then $n-2$, and so on. In total, this will be

$$n+(n-1)+(n-2)+...+2+1 = \frac{n(n+1)}{2} = 2n^2 + \frac{n}{2} = O(n^2)$$

So, we see that the obvious way to sort a list takes quadratic time, which should suggest to you that there's a better way to do it.

A superior sorting method is called MergeSort. It works by dividing L into two equal-sized pieces, recursively MergeSorting them both, and then merging the two sorted lists into a single sorted list. It looks like the following:

```
Function MergeSort
Input:
    List L of length n
Initialization:
    NewList = Empty list
Algorithm:
    If len(L)=1: Return L
    Else:
        L1 = L[1:len(L)/2]
        L2 = L[len(L)/2:]
        S1 = MergeSort(L1)
        S2 = MergeSort(L2)
        Return Merge(M1, M2)
```

where the function Merge looks like the following:

```
Function Merge
Input:
    Lists L1, L2
Initialization:
    NewList = empty list
Algorithm:
    While ~L1.isEmpty() and ~L2.isEmpty():
        If (L1[0] <= L2[0]) or L2 is empty:
            mn = L1[0]
            Delete L1[0]
```

```
Else:
    mn = L2[0]
    Delete L2[0]
NewList.append(mn)
If ~L1.isEmpty(): NewList.extend(L1)
Elif ~L2.isEmpty(): NewList.extend(L2)
Return NewList
```

The Merge function will be O(*n*), if *n* is the length of its longest input. This means that if $T(n)$ is the time to MergeSort a list of length *n*, then

$$T(n) = O(n) + T\left(\frac{n}{2}\right) + T\left(\frac{n}{2}\right)$$

$$T(n) = O(n) + 2 * T\left(\frac{n}{2}\right)$$

$$T(n) = O(n) + 2 * \left(O\left(\frac{n}{2}\right) + T\left(\frac{n}{4}\right)\right)$$

$$T(n) = O(n) + O(n) + 2 * T\left(\frac{n}{4}\right)$$

$$T(n) = 2 * O(n) + 2 * T\left(\frac{n}{4}\right)$$

$$T(n) = k * O(n) + 2^k * T\left(\frac{n}{2^k}\right)$$

We can break this down until $n/2^k$ is about 1, that is, $k \sim \log_2(n)$. That will give us

$$T(n) = \log_2(n) * O(n) + 2^{\log_2(n)} * O(1) = O(n * \log(n))$$

21.4 Amortized Performance and Average Performance

One limitation of Big-O complexity is that it is worst-case performance. In the real world though, there are situations where the Big-O complexity of an algorithm is quite bad, but the observed runtime of the code scales well.

The simplest case of this is certain algorithms that are deliberately randomized, in a way that *could* run slowly but will almost certainly run fast. The QuickSort algorithm is the most famous example of this, and it is typically an even better way to sort a list than the MergeSort I showed you earlier. It also has $O(n^*\log(n))$ runtime, and the algorithm is

```
Function QuickSort
Input:
    List L of length n
Initialization:
    NewList = Empty list
Algorithm:
    elem = random element of L
    lessthan = [x for x in L if x<elem]
    morethan = [x for x in L if x>elem]
    sortedless = QuickSort(lessthan)
    sortedmore = QuickSort(morethan)
    return sortedless + [elem] + sortedmore
```

This algorithm will be $O(n^2)$ if, by random chance, whenever QuickSort is called (including its recursive calls), it picks elem to be the largest value in the list. QuickSort is $O(n^2)$ worst-case, but $O(n^*\log(n))$ on average.

The other big caveat to Big-O notation is called "amortized analysis." This comes into play if you are doing many operations that modify a data structure in a running program. Often we cannot design it so that all operations take $O(1)$ time, but we can make it so that over the course of time, they average out to be linear.

A great example of amortized performance is Python dictionaries. Usually, adding a new element to a dictionary is an $O(1)$ operation: it doesn't depend on how many elements are already in the dictionary. As more and more elements are added though, the dictionary's internal structure (which I'll describe later – it's called a "hash map") starts to fill up, and occasionally, the data must be reshuffled to make more room. Reshuffling a dictionary that contains n elements is an $O(n)$ operation, so as the dictionary grows, these reshufflings become more and more costly. However, they also become more rare, so that the average time taken over many changes is $O(1)$.

This script and the figure it generates demonstrate how this works. We start off with an empty dictionary and then add elements to it, one at a time, until there are 10 million of them. The time taken for each addition is recorded, and then we plot them all out. Most additions take a fixed, tiny amount of time. But occasionally, an addition will trigger a reshuffling of the data in the dictionary, and the time spikes up.

```python
import time
import matplotlib.pyplot as plt
times, d = [], {}
for i in range(10000000):
    start = time.time()
    d[i] = i
    stop = time.time()
    times.append(stop-start)
plt.plot(times)
plt.xlabel("dictionary size")
plt.ylabel("time to add an element (seconds)")
plt.show()
```

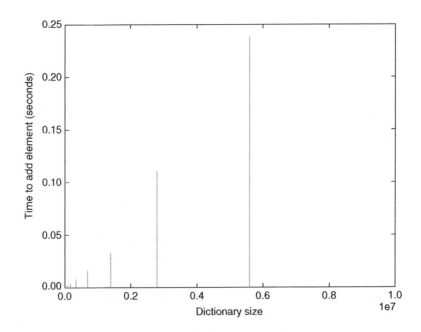

21.5 Two Principles: Reducing Overhead and Managing Memory

Big-O complexity tells you about how to make sure that your code scales and occasionally helps you avoid making catastrophically bad design choices. Now let's move on to the more mundane considerations, which involve the computer you're working on and the tools you're using. These won't change your code's Big-O complexity, but they could easily determine whether your code runs fast enough to be of any practical use.

The rest of this chapter will discuss several concrete tips for improving the performance of your code. However, those tips all fall under two broad categories, and I would like to briefly review them:

- Reducing overhead. Many operations that you do incur a certain amount of overhead every time you do them. These little performance penalties add up, especially if you are incurring them over and over again within a loop.
- Making better use of the computer's memory and caching.

The idea of reducing overhead is pretty self-explanatory. Understanding how to hack your computer's memory is a bit more complicated and deserves some explanation. This is especially the case because, in my experience, memory issues are more likely to be a big deal in data science than overhead is.

Computer memory is also called RAM or "random access memory." This terminology doesn't mean that there is anything actually random about it. Instead, the idea is that the computer can access any region of memory equally quickly. The time required varies widely by machine, but think on the order of 100 nanoseconds to move 1 byte of data from RAM to somewhere it can be processed.

Your computer only has a finite amount of RAM. If you force the computer to operate on more data than it can fit into physical memory, two things might happen: either your program will fail or your operating system will scramble to more data between RAM and the data storage medium (such as a hard disk) – this is an extremely time-consuming operation called "paging." So, one of the first principles of managing memory is to not take up too much of it.

Within the RAM, memory is not all equal. The computer will store a copy of part of RAM in what's called the cache, a piece of memory that has much faster read/write time (around an order of magnitude faster). When you try to read a byte of RAM, the computer first looks in the cache. If that byte is stored there, it just reads the copy, only resorting to the actual RAM in the case of a "cache miss." Similarly, if the program needs to modify a byte, it will first look for a cached copy of that byte and only modify that if it finds one. The computer will periodically write changes from the cache back to RAM in a batch process.

Actually, there are usually several levels of cache, each one smaller and more rapid access than the one below it. The runtime of a program will often be dominated by how often the processor can find that data it's looking for in the top levels of the cache.

Together, the RAM, disk, and various cache levels form the "memory hierarchy," where each layer is faster but smaller than the ones below it. Every time the computer needs a piece of data, it will look for it in the top level of the hierarchy, then the one below it, and so on, until the data is found. If a piece of data is located far down the hierarchy, then accessing it can be excruciatingly slow – partly because the access is inherently slow, but also because we just wasted time looking for the data higher up in the hierarchy. For this reason, how often the data can be found in the high levels is often the single most important contributor to a program's performance.

21.6   Performance Tip: Use Numerical Libraries When Applicable

Operating on NumPy arrays (directly or through a library such as Pandas) is a lot more efficient than operating on Python objects. The following code looks at the time required to increment all elements of a list of numbers, for a Python list and for a NumPy array. It plots the Python/NumPy ratio as a function of the list length, and you can see that the ratio starts high and gets worse.

There are two big factors that contribute to this:

- It takes longer to add 1 to a number in pure Python than in NumPy. This is because Python doesn't know until runtime that the number is an integer. It must check for each number what its data type is and hence how to compute "$x + 1$." This additional overhead will make Python a constant factor slower than NumPy.
- The Python data structure takes up a lot more space compared to the NumPy array, because it must also carry around metadata that specifies what data type each list element is. All of this extra memory means that you fill up the high levels of cache faster in Python than in NumPy, meaning that Python does comparatively worse and worse for longer lists.

```
import time, numpy as np, matplotlib.pyplot as plt
def time_numpy(n):
    a = np.arange(n)
    start = time.time()
    bigger = a + 1
    stop = time.time()
    return stop - start
def time_python(n):
    l = range(n)
    start = time.time()
    bigger = [x+1 for x in l]
    stop = time.time()
    return stop - start
n_trials = 10
ns = range(20, 500)
ratios = []
for n in ns:
    python_total = sum([time_python(n)
```

```
        for _ in range(n_trials)])
    numpy_total = sum([time_numpy(n)
        for _ in range(n_trials)])
    ratios.append(python_total / numpy_total)
plt.plot(ns, ratios)
plt.xlabel("Length of List / Array")
plt.ylabel("Python / Numpy Ratio")
plt.title("Relative Speed of Numpy vs Pure Python")
plt.show()
```

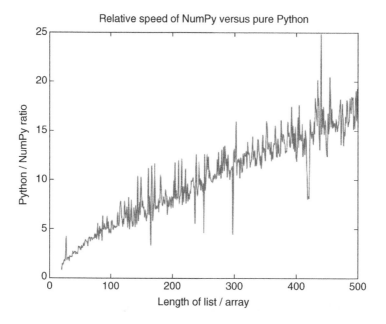

21.7   Performance Tip: Delete Large Structures You Don't Need

At any point in the running of your code, all objects that you have created will be vying for space in the cache. If you get rid of the ones that are no longer needed, then you allow for the ones that you will end up needing to be higher up in the memory hierarchy.

The process of deleting data structures that you no longer need, and hence freeing up the memory that they were taking up, is called "garbage collection." Python will do a certain amount of it automatically for you. For example, after a function finishes running all of the variables that were defined in it that can no longer be accessed are flagged for eventual deletion.

However, Python is very conservative about when it deletes data structures, so you can free up a lot of space by deleting objects manually.

This is done using the "del" keyword, as follows:

```
>>> del my_object
```

21.8   Performance Tip: Use Built-In Functions When Possible

Python's built-in functions are written in very efficient C code, and calling one only incurs the one-time performance hit that is inherent in calling any function. The following code compares the time required to add up all of the numbers in a list using the sum() function versus the time to add them up using a loop. On my computer, the latter takes about 14 times as long.

```
l = range(10000000)
start = time.time()
_ = sum(l)
stop = time.time()
time_fast = stop - start
start = time.time()
sm = 0.0
for x in l: sm += x
stop = time.time()
time_loop = stop - start
print "The ratio is", time_loop / time_fast
```

21.9   Performance Tip: Avoid Superfluous Function Calls

Every time a Python function is called, there is a certain amount of overhead. The following code compares the time to loop over a list and add up all of its values, versus looping over it and calling a function that adds the values. Doing the latter takes (on my computer) about twice as long: the overhead required to just call a function is equal to the time required to move to the next iteration of a loop, plus the time to actually perform the addition.

```
add_nums = lambda a, b: a+b
l = range(10000000)
start = time.time()
sm = 0
```

```
for x in l: sm += x
stop = time.time()
time_fast = stop - start
start = time.time()
sm = 0
for x in l: sm = add_nums(sm, x)
stop = time.time()
time_func = stop - start
print "The ratio is", time_func / time_fast
```

21.10   Performance Tip: Avoid Creating Large New Objects

A common performance mistake is to create large new objects from existing ones, when they could instead have been updated. For example, saying

```
>>> myList = myList + [1, 2, 3]
```
and
```
>>> myList.extend([1,2,3])
```

will have the same effect. However, the first one will involve creating a new list, copying all of the contents of myList into it, and deleting the old myList. This is an $O(n)$ operation! The second one is amortized $O(1)$, and I've seen to turn computations that always failed to ones that went through quickly.

21.11   Further Reading

1 Scott, M, *Programming Language Pragmatics*, 4th edn, 2015, Morgan Kaufmann, Burlington, MA.
2 Petzold, C, *Code: the Hidden Language of Computer Hardware and Software*, 2000, Microsoft Press, Redmond, WA.

21.12   Glossary

Amortized complexity The average $O(1)$ performance of an operation if it is done many times. Sometimes, it maybe be $O(n)$ or even worse, but those are rare enough that it is $O(1)$ on average.
Big-O notation An algorithm is $O(f(n))$ if its asymptotic performance as n gets big is upper-bounded by $f(n)$, times some constant factor.
Cache A low-latency piece of memory where certain data in RAM can be stored for more rapid access.

Cache miss When a program looks in the cache for a piece of data but doesn't find it. The program must then do the much more expensive operation of reading from normal RAM.

Complexity The asymptotic performance of an algorithm as measured in Big-O notation.

Garbage collection Deleting data structures that are no longer needed from memory, which frees up space.

Iterator A programming abstraction that provides values one at a time. Often, only a few values are ever actually held in memory at once, so iterators make exceptionally good use of caching.

Part III

Specialized or Advanced Topics

The rest of this book will cover several advanced topics. Some of the things we discuss really are advanced topics, which are often useful in data science but which many data scientists never need. Deep learning is a good example of this.

In other cases though, we will be fleshing out topics that we've already discussed. There will be less in the way of nuts-and-bolts code and more abstract theory. The big reason for this is that standard techniques often don't work for one reason or another. For example, you might need to adjust how a machine learning model works in order to accommodate outliers in a particular way. If this happens, you will have to revisit the assumptions that the standard techniques are based on and devise new techniques that work for your situation.

The Data Science Handbook, First Edition. Field Cady.
© 2017 John Wiley & Sons, Inc. Published 2017 by John Wiley & Sons, Inc.

22

Computer Memory and Data Structures

This chapter dovetails with the one on computer performance. It will describe in more detail the way a computer program's memory is laid out and how data is encoded in memory. It will then move on to some of the most important data structures in common use and explain how their physical layout gives rise to their performance characteristics.

To make things concrete, this chapter will be taught using the C language. C is a low-level language that gives you very fine-grained control over how a program utilizes memory. The main Python interpreter, similar to a lot of the most important code in the world, is written in C because it allows you to make things very, very efficient. This chapter doesn't count as a crash course in C, but it will give you enough to understand how the key data structures are implemented and how they form the basis of Python.

22.1   Virtual Memory, the Stack, and the Heap

One of the most important jobs of an operating system is to allow multiple different processes on the computer to share the same physical RAM. It does this by providing each process with a "virtual address space" (VAS), which it can use to store the data it is operating on. The process can refer to data in any location from 0 to 2^{32}-1 in 32-bit operating systems and 0 to 2^{64}-1 in 64-bit operating systems. Each location contains exactly 1 byte of data, and the finite range of valid addresses puts a hard (but very large) upper limit on the amount of data that the process can be operating on at once. The operating system takes care of which addresses in the VAS correspond to which locations in physical RAM. It may also shift this mapping around, moving data between RAM, the different layers of the cache, and the long-term disk memory. From your program's perspective though, the VAS is all there is.

The process cannot just access data in the VAS willy-nilly. It must first request that the operating system set aside some physical RAM and match it up with addresses in the VAS. If the process ever tries to access a location in the

The Data Science Handbook, First Edition. Field Cady.
© 2017 John Wiley & Sons, Inc. Published 2017 by John Wiley & Sons, Inc.

VAS that the OS hasn't allocated to it, this is called a "segmentation fault" or "seg fault." Seg faults are a notorious class of bugs that are a constant headache for people who use low-level languages such as C, but that are mercifully impossible in a language such as Python. When space in the VAS is no longer needed, the process should "free" the memory, notifying the OS that that range of addresses is no longer needed and it can allocate the physical RAM to another process.

22.2 Example C Program

Ok, now that you know how memory is laid out, let's dive into some C code. Here is the famed "hello, world" program – it just prints "hello, world" on the screen, but it shows you the essential parts of a C program:

```
#include <stdio.h>
#include <stdlib.h>
int main(void) {
  printf("hello, world\n");
  return 0;
}
```

The most important thing about a C program is the presence of the subroutine called "main." As the name suggests, this will be the primary routine that the program will start off with running. Having a function called "main" is optional in Python, and there is nothing magical about "main"; in C, this function is an absolute necessity. We say "int main" because the main function returns the integer 0, which gets passed to the operating system when our program finishes. The first line is saying to include the stdio library, which contains the "printf" function for printing to the screen.

There are several ways to compile and run this code. On my computer (a mac, with the developer toolkits installed), it looks like the following:

```
$ # Assume the code is in a file called mycode.c
$ gcc mycode.c
$ # a.out is the default name of the executable file
$ ./a.out
hello, world
```

22.3 Data Types and Arrays in Memory

The previous code just shows you the basic syntax of a C program, which is nice to know but which I don't want to dwell on. The really interesting thing is the way that C defines its types.

Let's dive into a more interesting piece of code and see what it's up to:

```c
#include <stdio.h>
#include <stdlib.h>

int main() {
 // chars, i.e. single characters
 char mc = 'A';
 printf("The char is %c\n", mc);
 // doubles, i.e. floating numbers
 double d = 5.7;
 printf("The double is %d\n", d);
 // arrays of ints
 int myArray = {1,3,7,9};
 for (int i=0; i<4; i++) {
  printf("The %ith number is %i\n", i, myArray[i]);
 }
return 0;
}
```

There are two big things going on here. First, there are data types other than integers floating around. There's a "double," which is used to hold decimal numbers such as 5.9. There is also a "char," which is short for character, and it can hold a single 1-byte character such as a letter or a digit.

Each of these atomic data types takes up a fixed number of bytes in memory. How many bytes that is will depend on your computer – my computer uses 4 bytes for an int – but the key thing is that the size is fixed.

The myArray variable is what's called an "array of ints," and you can have arrays of any other fixed-size data type too. Under the hood, the bytes of the array will just be the bytes of its constituent integers all concatenated together.

In this code, we use a for-loop to loop over the array and print out all of its values. The line

```c
    for (int i=0; i<4; i++)
```

is the way that C does for-loops. It means that we should have an integer called i, which we use for the index of the loop. i will be initialized to 0 and will be incremented for every new loop. When i is no longer less than 4, the loop will terminate.

The fact that a data type takes up a fixed number of bytes is at the corner of how arrays operate. Because integers all take up the same number of bytes, and myArray is just those bytes concatenated together, it is very easy to pull out the ith element from myArray. It will just be the group of bytes that it offset by i*(number of bytes in an int) from the start of myArray. This is the way that the computer can tell where one int ends and another begins, and it is an O(1) operation to take out the ith element regardless of the size of the array.

When you use NumPy, all of the numbers are stored as C arrays under the hood. This is why they take up so little space and are so quick to operate on: it's ultimately just C for-loops that operate on the raw bytes. This is also why NumPy arrays, unlike Python lists, are constrained to only contain data of the same type; all elements must be of the same size, and a particular logical operation on them must correspond to the same operation of the underlying bytes.

22.4 Structs

Now that you know about atomic, fixed-sized types and arrays of them, let's look at the simplest compound data type: the struct. Here is some example code:

```
#include <stdio.h>
#include <stdlib.h>
struct Person {
   int age;
   char gender;
   double height;
};
typedef struct Person Person;
int main() {
   Person Bob = {30, 'M', 5.9};
   printf("Bob is %f feet tall\n", Bob.height);
   return 0;
}
```

Look at the following code:

```
struct Person {
   int age;
   char gender;
   double height;
};
typedef struct Person Person;
```

The first part defines the new data type, called a "struct Person." A struct Person contains an integer called age, a char called gender, and a double called height. It's pretty ugly though to write "struct Person" in two words when it's really just a single thing. So, purely for notational convenience, the "typedef" line defines the term "Person" to be equivalent to "struct Person."

To picture what's going on in the code, we will often draw the Person in this code as follows:

Bob	
Age	30
Gender	M
Height	5.9

Similar to the atomic datatypes, a Person will take up a fixed number of bytes. The bytes for a particular Person will just be the bytes for their age, gender, and height, all concatenated together into a larger array of contiguous bytes. The fields will be in an order that is chosen by the compiler, possibly with some spacer bytes thrown in for performance reasons. Compiled code won't make reference to the "gender" field of a Person – it will just talk about the bytes that are offset a certain distance from the start of the Person.

Similar to the atomic types, we can have arrays of structs in memory, and we can access their data efficiently. If you have a long array of Persons and you want the gender of the nth Person, then it will simply be the byte that is offset by n*(number of bytes for a Person) + (offset for the gender field) from the start of the array.

22.5 Pointers, the Stack, and the Heap

There is a very, VERY important data type that I did not mention in the previous section: the pointer. A pointer is stored as an integer (i.e., it is fixed-size), but it stores the index in the VAS of a particular byte of memory. This allows us to refer to data in arbitrary location in the VAS, which may be the start of an object or array of arbitrary size and complexity.

Let's look at a more complicated version of the previous code.

```
#include <stdio.h>
#include <stdlib.h>
struct Person {
   int age;
   char gender;
   double height;
   struct Person* spouse;
};
typedef struct Person Person;
void marry(Person* p1, Person* p2) {
   p1->spouse = p2;
   p2->spouse = p1;
}
```

```
int main() {
    Person Jane = {28, 'F', 5.5, NULL};
    Person Bob = {30, 'M', 5.9, NULL};
    marry(&Jane, &Bob);
    printf("Janes spouse is %f feet tall\n",
        Bob.spouse->height);
    printf("Bob is %f feet tall\n", Bob.height);
    return 0;
}
```

The first change to notice is that the Person struct now has new field called "spouse," of type Person.* The Person* data type is not a Person, but rather a pointer to some byte in memory, which is to be interpreted as the first byte in a Person. This way a Person can contain not only data about a single human being but also references to other human beings.

The special term NULL means that the pointer does not actually point to anything; under the hood, it is stored as all 0s. When we first initialize our structures, we will draw them as follows:

Bob	
Age	30
Gender	M
Height	5.9
Spouse	

Jane	
Age	28
Gender	F
Height	5.5
Spouse	

The next line of code calls the "marry" function, which we can see takes in two Person* as its arguments. "&Bob" is special syntax that means a pointer to the Field structure, and similarly for &Jane, so we are giving it what it wants. The one thing left to note is that we use "->" rather than "." to access the member of a structure that we have a reference to. Passing the function a pointer to our structure is called "passing by reference."

The result of the marry() function will be to modify the two objects in memory so that they look like as follows:

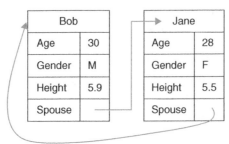

When we passed Bob and Jane into the marry() function, we did what's called "passing by reference": we passed pointers to Bob and Jane into the marry() subroutine, so that any changes marry() makes will happen to the original data structures in memory. When marry() finishes, the changes that it made will persist.

This might be ringing a bell for you. In Python, objects are *always* passed by reference. This means that if you pass around a mutable object such as a list or dictionary, it can be changed as follows:

```
>>> A = {1:1, 2:4, 3:9}
>>> B = A  # pointer to the same dict
>>> B[4] = 16
>>> A[4]
16
```

The alternative to passing by reference is "passing by value": whenever you want to pass a data structure into a subroutine, make a copy of the data structure and put it in some place that the subroutine knows where to look. There are situations where this is more performant, because the computer no longer needs to waste time following pointers and fetching the data they point to. However, this incurs the computational cost of copying the entire (potentially massive) data structure over every time you want to pass it into a subroutine.

The overwhelming reason to generally use the heap rather than the stack is that, for technical reasons involving the way subroutines work, the compiler has to know how big all data structures are at compile time. This is the case in the previous code, since we only created two Person structures.

But if we wanted to make an array of Persons, and have a user tell us how many there should be every time the code gets run, then that array has to be put into the heap. This involves requesting from the operating system, at runtime, that it allocate a block of heap space of a given size. As the program runs, it will also involve telling the OS when the memory is no longer needed.

Let's jump into some more code, giving the Person struct a pointer to the array of its children. The malloc() function is used to request space in the heap – it allocates the memory somewhere in the heap and returns a pointer to the first byte.

```
struct Person {
    int age;
    char gender;
    double height;
    struct Person* spouse;
    int n_children;
    struct Person* children;
};
```

```
typedef struct Person Person;
void marry(Person* p1, Person* p2) {
  p1->spouse = p2;
  p2->spouse = p1;
}
int main() {
  Person Jane = {30, 'M', 5.9, NULL};
  Person Bob = {28, 'F', 5.5, NULL};
  marry(&Jane, &Bob);
  printf("Jane is %f feet tall\n", Jane.height);
  printf("Bobs spouse is %f feet tall\n",
    Bob.spouse->height);
  int NUM_KIDS = 5;
  Jane.n_children = NUM_KIDS;
  Jane.children = (Person*) malloc(
     NUM_KIDS*sizeof(Person));
  Bob.n_children = NUM_KIDS;
  Bob.children = Jane.children;
  for (int i=0; i<NUM_KIDS; i++) {
   Person* ith_kid = &Jane.children[i];
   if (i<3) ith_kid->gender='M';
   else ith_kid->gender='F';
  }
  int n_sons = 0;
  for (int i=0; i<NUM_KIDS; i++) {
   if (Jane.children[i].gender=='M') n_sons++;
  }
  printf("Jane has %i sons\n", n_sons);
  free(Jane.children);
  return 0;
}
```

Now a Person contains not just Person* spouse but also Person* children. Children will point to an array of Person structs in memory. It can be an arbitrarily short or long array, so it will have to be located in the heap. Note that when you look at the bytes themselves, there will be no way to tell when the array of Persons ends, so we will also have to keep track of how long the array is, so we need the int n_children to keep track.

Most of the rest of the code should now make sense, with the exception of "free(Jane.children)." This is the way we notify the operating system that the range of memory in the VAS is no longer needed.

If you don't free up the memory in the heap, it's called a "memory leak," and it's a surprisingly easy bug to write. As your program runs, it will constantly allocate new variables in the heap, forgetting that they are there, until all of the memory has been taken up and the program crashes. The following code looks like it should run forever, but it will eventually fail when all the heap space is taken up:

```
Person* bob;
while(true) {
// creates new Person in the heap and points bob at it
// however, does not free up the previous Person
bob = malloc(sizeof(Person));
}
```

Memory that has been allocated in the heap but that there is no longer a pointer to is called "orphaned."

Before we move on, I should note that my code here has been extremely sloppy. I wrote it to make it easy to understand, but please don't write professional code such as this! In particular, I've been very cavalier about whether pointers were pointing to valid places in memory. When I allocated the children, for example, I should have made sure that all of their own spouse and children pointers were either NULL or pointing at valid locations. Again, you generally won't have to worry about this stuff in your own coding, since a language such as Python takes care of all this boilerplate for you. But if you want to understand how Python works, this is what's going on under the hood.

22.6 Key Data Structures

The previous sections introduced the key concepts of fixed-size structs, arrays of structs, and pointers. This section will show how those ingredients can be mixed together to create a wide range of complex, efficient data structures. I will only show you a few of them, but these structures are the basis of Python and every other piece of software you use.

22.6.1 Strings

Given that strings are an atomic type in Python, it might seem counterintuitive that they are not one in C. This is because strings are of variable length, so they must be created on the heap and the program must keep track of how long they are. Strings are universally stored as a char* – a pointer to an array of bytes that are interpreted as characters (usually with ASCII encoding).

There are two main ways to keep track of how long the array of characters is:

- These days it is very common to wrap the char* in a structure that keeps track of how many characters there are and possibly other information. A simple example might look like this:

```
struct MyString {
  char* characters;
  int n_chars;
};
```

- To save space, in the past it was common to signify the end of a char* by having the last byte be 0. Such a char* is called "null-terminated". This approach is trickier to work with and takes more time to process. However, it does take up a little less RAM and used to be a lot more important than it is now.

22.6.2 Adjustable-Size Arrays

The problem with an array in memory is that, while it can modify its elements, its size is fixed. Adding a new element will require allocating a new array of sufficient size, copying the original array over, and putting the last element in place. This is an $O(n)$ operation!

Adjustable-size arrays turn it into an amortized $O(1)$ operation. The idea is to allocate more space in the array than needed and add new elements into the extra spaces whenever they are appended. The C code might look like this:

```
struct AdjustableList {
  int array_size;
  int n_chars;
  char* characters;
};
```

and the layout of memory might be as follows:

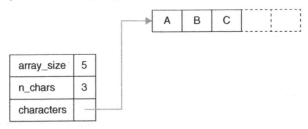

When the allocated array does finally fill up, then yes, we will have to make a new one and copy the first one's contents over at high cost. But a common way to do this is to double the size of the array every time. In that case, the copy operations become twice as expensive as the list grows, but half as frequent. This makes a net $O(1)$ cost amortized among all of the additions.

The Python list of object is, under the hood, and adjustable-size array. The difference though is that it isn't an array of ints or anything – it is an array of pointers, which point to arbitrary Python objects. So, the code

```
myList = [1, 2,    []]
```

would result in memory layout that looks something like this:

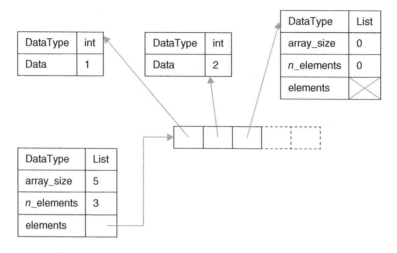

22.6.3 Hash Tables

The key limitation of arrays is that you can only index them by integers. Really, you can think of an array as a map from integers to values. In other cases though, you want to have a map that takes in a more flexible type, such as a string. This is made possible by a data structure called a hash table.

The key feature of a hash table is something called a hash function. Hash functions are ubiquitous in computing, so they're important to understand. A hash function $f()$ takes in an object x of whatever input type is desired, and it outputs an integer that falls within some range from 0 to N. The hash function f has three important properties:

- It is deterministic, so that it will always return the same integer, given the same input.
- It takes $O(1)$ time to compute $f(x)$ for any x.
- f "looks random." For a random x, $f(x)$ is about equally likely to be any of the integers. If $x != y$, then almost certainly $f(x) != f(y)$.

Basically, a hash function is a way to deterministically garble its input into a single number.

Let's say we have a key/value pair that we want to store. The key idea of a hash table is that we still store our data in an array, but the index in the array is given by the hash of the key. This means that looking up an element is still $O(1)$ time, since calculating the hash function is $O(1)$. Typically in a hash table, the underlying array is an array of pointers, and a pointer will be NULL if there is no element that got hashed to that location in the array. As with the aforementioned adjustable-size arrays, the pointers can point to objects of arbitrary and perhaps differing sizes.

Hash tables play a central role in Python. Dictionaries are implemented as hash tables under the hood and so is the namespace, which maps your variable names to the objects in memory that contain the data. Ditto for instances of user-defined classes that you create. It has been said that the entire Python language is just syntactic sugar around hash tables, and it's really pretty true.

It will periodically happen for two distinct x and y, and we will have $hash(x) = hash(y)$. This inconvenience leads to two caveats in practical implementations of hash tables:

- Generally, we can't just store the values in the hash table. We must show the key and the values at the location in the array, so that the raw keys themselves can be compared.
- You could try to hash a key x to a location in the array that already has a different key y. Sometimes, this is solved by having a protocol for how to look around and find an empty cell in the array. Other times, the cells in the array point to yet another structure, whose purpose is to keep track of whatever values mapped to that cell.

At some point, a hash table starts to fill up, so that most cells in the underlying array are holding many different values. At this point, we must go through a very expensive operation called "rehashing." We allocate a new, larger hash table in memory. It will need a new hash function, which maps values into a larger range of integers. Then, we go through all of the key/value pairs in the original table and put them into the new, larger one. This is an $O(n)$ operation, but if you increase the array size by a constant factor every time, then adding new elements becomes amortized $O(1)$.

22.6.4 Linked Lists

A linked list is a lightweight way to implement a list of objects of some known type. You have a struct that has two fields: an instance of whatever it is that you're making a list of and a pointer to the next element in the list. The last element in the list points to NULL.

For example, here is some potential code for a linked list of integers:

```
struct LinkedListNode {
  int Value;
  LinkedListNode* next;
};
```

Here is what the list [5, 4, 7, 8] would look like when store using it:

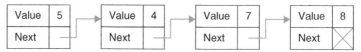

Linked lists have the huge disadvantage that it is an O(*i*) operation to find the ith element, rather than O(1). However, the big advantage of linked lists (besides their simplicity) is that you can add a new element to them in O(1) time, if you have a pointer to the location where you want to put it. This is pure O(1) time, not amortized.

The pseudocode looks like this:

```
Input: LinkedListNode* currentNode, int n
Algorithm:
   LinkedListNode* newNode = new LinkedListNode(n)
   newNode->next = currentNode->next
   currentNode->next = newNode
```

and the operation in memory will look something like this:

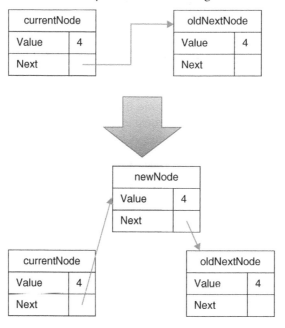

Note that the second line of the algorithm MUST happen before the third. If we had switched the order, we would have pointed currentNode at newNode, and then pointed newNode back at itself, as follows:

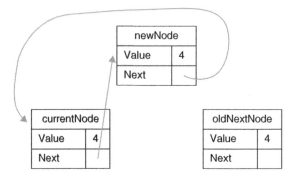

Now the list loops back on itself perversely, making it seem like it is infinitely long, and the remainder of the original list has been orphaned and is floating round in memory.

It is notoriously easy to screw up pointer manipulations, and that is one of the things that make high-level languages such as Python very appealing. On the other hand, if you're willing to put in the effort to get all the details right, then C code can achieve performance that Python can only dream of.

22.6.5 Binary Search Trees

There is one other pointer-based data structure I would like to introduce you to: the binary search tree (BST). In a BST, every node contains a numerical value and has two children rather than one, which we typically call "left" and "right." We make sure that at every point in time, the children on the right of every node have values that are greater than or equal to the node in question, and the nodes on the left have smaller or equal values. A BST might look like this:

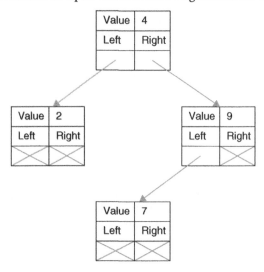

Assuming that every node in the tree has roughly as many descendents to its left as to its right (a so-called balanced tree), then it is an O(log(*n*)) operation to see whether a given number is in the tree or add a new one. Again, this is not amortized performance. The search algorithm, to find whether a value is in the BST, would look like this in pseudocode:

```
BST_Search
Input: Node* currentNode, int n
Algorithm:
  If currentNode==NULL: return False
  Elif n < currentNode->value:
    return BST_Search(currentNode->left, n)
  Else:
    Return BST_Search(currentNode->right, n)
```

There is so much more to say about pointers, arrays, and the magic that you can do with them. I would love to get into it, but that's partly because I spent many years cutting my teeth on low-level coding. In the daily practice of a data scientist, you don't need to use this information, so use this chapter as a primer in case you ever find yourself writing low-level algorithms.

22.7   Further Reading

1 Petzold, C, *Code: the Hidden Language of Computer Hardware and Software*, 2000, Microsoft Press, Redmond, WA.
2 Scott, M, *Programming Language Pragmatics*, 4th edn, 2015, Morgan Kaufmann, Burlington, MA.
3 McDowell, G, *Cracking the Coding Interview: 189 Programming Questions and Solutions*, 6th edn, 2015, CareerCup.

22.8   Glossary

Binary tree   A data structure where each node has pointers to two children, on its left and right.
Hash table   A data structure that maps hashable keys to values. It does that by hashing the keys into the range $[0, N]$ and using the hash as an index in a length-N array.
Hash function   A deterministic function that garbles a key x into an integer that is probably distinct from hash(y), for $y != x$.
Heap   A range in the virtual address space where dynamically allocated memory is located.

Linked list A data structure where each node has a pointer to the next node. The last node has a NULL pointer.

Orphaned memory Memory in the VAS that there is no longer a pointer to. This makes it impossible to free the memory.

Memory leak Part of a program that clears pointers to memory in the heap without freeing the memory.

Passing by value Passing a copy of a data structure into a subroutine. That way changes made to the copy do not to effect the original version.

Passing by reference Passing a pointer to a data structure into a subroutine. That way changes made to the object do effect the original object.

Pointer The address of a location in the virtual address space. That address marks the first byte in the data object being pointed to.

Stack A range in the virtual address space that is involved in low-level subroutines.

Struct A data type that combines several data fields of more primitive, fixed-size types. The struct is itself fixed-size, and its byte representation is just the concatenated byte representation of its constituent data fields.

Virtual address space The array of bytes that a computer process has for storing all of its data that it operates on. The operating system handles the mapping between the logical addresses and their physical location in RAM.

VAS Short for Virtual Address Space.

23

Maximum Likelihood Estimation and Optimization

This section will talk about two topics that form the mathematical and computational underpinnings of much of what we've covered in this book. The goal is to help you frame novel problems in a way that makes theoretical sense and that can realistically be solved with a computer.

23.1 Maximum Likelihood Estimation

Maximum likelihood estimation (MLE) is a very general way to frame a large class of problems in data science:

- You have a probability distribution characterized by some parameters that we'll call θ. In a regular normal distribution, for example, θ would consist of just two numbers: the mean and the standard deviation.
- You assume that a real-world process is described by a probability distribution from this family, but you do not make any assumptions about θ.
- You have a dataset called X that is drawn from the real-world process.
- You find the θ that maximizes the probability $P(X|\theta)$.

A large fraction of machine learning classification and regression models all fall under this umbrella. They differ widely in the functional form they assume, but they all assume one at least implicitly. Mathematically, the process of "training the model" really reduces to calculating θ.

In MLE problems, we almost always assume that the different data points in X are independent of each other. That is, if there are N data points, then we assume

$$P(X \mid \theta) = \prod_{i=1}^{N} P(X_i \mid \theta)$$

In practice, it is often easier to find θ that maximizes the log of the probability, rather than the probability itself. Taking the log turns multiplication into addition, so that we are minimizing

The Data Science Handbook, First Edition. Field Cady.
© 2017 John Wiley & Sons, Inc. Published 2017 by John Wiley & Sons, Inc.

$$\log\big(P(X\,|\,\theta)\big) = \sum_{i=1}^{N} \log\big(P(X_i\,|\,\theta)\big)$$

There are two significant problems with MLE in general. The first is overfitting; especially if your dataset is small, you run the very real risk of getting parameters that predict your specific data very well, but generalize terribly. The most popular alternative to this is a Bayesian approach, which is generally much harder to understand, more complicated to implement, and slower to run. Plus, Bayesian approaches involve the very touchy-feely issue of how you pick your prior.

The other problem with MLE is the logistical problem of actually calculating the optimal θ. In some cases, there is a tidy closed-form solution – those cases tend to be the most historically important ones, if only because people could actually solve the problems back in the days of paper and pencil. In general though, there is no closed-form solution to an MLE problem, and we must rely on numerical algorithms that give us good approximations. This falls under the umbrella of numerical optimization, the other topic of this chapter.

23.2 A Simple Example: Fitting a Line

To illustrate, let's reproduce simple least-squares line fitting in the MLE context. In this case, there is, luckily, a closed-form solution that we can derive with some algebra and calculus.

Let's assume that y is a linear function of x plus some random noise, and let the noise be normally distributed. Then we see that the probability density of Y, for a given value of x, is

$$f_Y(y) = \exp\left\{ -\frac{\big((mx+b)-y\big)^2}{2\sigma^2} \right\}$$

The MLE expression we want to minimize is

$$L = \log\big(P(X\,|\,\theta)\big)$$
$$= \sum_{i=1}^{N} \log\big(P(X_i\,|\,\theta)\big)$$
$$= \sum_{i=1}^{N} \log\big(P(X_i\,|\,m,b,\mu,\sigma)\big)$$
$$= \sum_{i=1}^{N} \log\left(\exp\left\{ -\frac{\big((mx_i+b)-y_i\big)^2}{2\sigma^2} \right\} \right)$$

$$= \sum_{i=1}^{N} - \frac{\left(\left(mx_i + b\right) - y_i\right)^2}{2\sigma^2}$$

$$= \frac{-1}{2\sigma^2} \sum_{i=1}^{N} \left(\left(mx_i + b\right) - y_i\right)^2$$

In order for this expression to be at its maximum, the derivates of L with respect to m and b must be 0. So, we say

$$0 = \frac{\partial L}{\partial m} = \sum_{i=1}^{N} \frac{\partial}{\partial m} \left(\left(mx_i + b\right) - y_i\right)^2 \propto \sum_{i=1}^{N} \left(\left(mx_i + b\right) - y_i\right) {}^* x_i$$

$$0 = \frac{\partial L}{\partial b} = \sum_{i=1}^{N} \frac{\partial}{\partial b} \left(\left(mx_i + b\right) - y_i\right)^2 \propto \sum_{i=1}^{N} \left(\left(mx_i + b\right) - y_i\right)$$

If we divide each side by N, these equations then become

$$0 = m\left(\frac{1}{N}\sum_{i=1}^{N}x_i^2\right) + b\bar{x} + \left(\frac{1}{N}\sum_{i=1}^{N}x_i y_i\right)$$

$$0 = m\bar{x} + b + \bar{y}$$

We can solve these equations to find the complicated (but closed-form!) solutions

$$m = \frac{\left(\dfrac{1}{N}\sum_{i=1}^{N}x_i y_i\right) - \bar{x} * \bar{y}}{\left(\dfrac{1}{N}\sum_{i=1}^{N}x_i^2\right) - \bar{x}^2}$$

$$b = -\bar{y} - m\bar{x}$$

This process illuminates plain old vanilla least squares. It is no longer just a standard black-box technique; it is the *right* technique if you make certain assumptions about the world. In particular, we assumed that y is a linear function of x plus some normally distributed noise. But is that really the correct assumption?

If x is somebody's income and y is the price of their home, then I might want to use a different model, where the noise term has standard deviation proportional to x. This is because a difference of $20k in home price is very significant if you make $50k a year, but small if you make $500k. In that case, I would want to use

$$L = \sum_{i=1}^{N} \log\left(\exp\left\{ -\frac{\left((mx_i + b) - y_i\right)^2}{2(\beta x_i)^2} \right\} \right)$$

where the standard deviation in home price will be β times a person's income. This expression probably doesn't have a closed-form solution, but you can get an approximate numerical one.

An alternative variation on least squares would be to use something other than a normal distribution. In real life, people sometimes buy houses that are well above or below their means, so we might want a distribution that allows more outliers than the normal distribution.

The sky is the limit with the variations you can take. But the bottom line is this: rather than blindly trusting standard techniques, MLE allows us to take an understanding of the real world, translate it into probabilistic models, and bake those directly into the model.

23.3 Another Example: Logistic Regression

A more complicated example is the logistic regression classifier. Recall that in logistic regression, x will be a vector and y will be either 0 or 1. The score that it gives out is

$$p(x) = \sigma(w \cdot x + b) = \frac{1}{1 + \exp(w \cdot x + b)}$$

where w is a vector and b is a constant.

Logistic regression is based on the probability model that $p(x)$ is the probability that a real-world point at x will be a 1, rather than a 0. Conversely, $1 - p(x)$ is the probability that it will be a 0. Given the overall set of x values in our training data, the likelihood of the particular y values that we saw is

$$L = \left(\prod_{y_i=1} p(x_i) \right)\left(\prod_{y_i=0} (1 - p(x_i)) \right)$$

The training phase in logistic regression finds the w and b that maximize this expression.

23.4 Optimization

Optimization is a way to solve the following problem:

You are given a function $f(\theta)$, where θ is a vector of d dimensions.
You might also be given several functions $g_i(\theta)$.
Find the θ that minimizes $f(\theta)$, consistent with $g_i(\theta) \le 0$ for all i.

In my experience, students go through three distinct stages when first learning about optimization:

- What the heck? The concept of optimization is so general as to be meaningless. Where's the beef?
- Oh my gosh, *everything* is an optimization problem! This is the solution to life, the universe, and everything!
- Ok, it turns out that lots of things aren't optimization problems. And most of the ones that are optimization problems in principle can't be solved in practice. But there are still a *lot* of problems that can.

Numerical optimization is the way that many MLE problems get solved in the real world, but it is useful in many other domains of application as well.

As a simple example, MLE is prone to overfitting, as we discussed previously. This can be ameliorated by adding in penalty terms that punish parameters that are likely to be overfitted. With logistic regression, for example, we can add in a term to punish large feature weights:

$$L = \left(\prod_{y_i=1} p(x_i) \right) \left(\prod_{y_i=0} (1 - p(x_i)) \right) - \lambda \sum_i |w_i|$$

where λ is a positive parameter that you set. This is no longer a valid MLE problem, but it still fits perfectly well under the umbrella of optimization. In fact, it is now LASSO regression.

By taking the absolute value of the feature weights, LASSO regression punishes small weights and large weights moderately. However, we might also want to punish large weights very harshly to make sure that no one feature tends to domain the classification. This can be done by adding a penalty term that squares the weights, rather than taking their absolute value:

$$L = \left(\prod_{y_i=1} p(x_i) \right) \left(\prod_{y_i=0} (1 - p(x_i)) \right) - \lambda_1 \sum_i |w_i| - \lambda_2 \sum_i |w_i|^2$$

This variant is known as elastic net regularization.

Most of the time as a data scientist, you can use off-the-shelf algorithms and just know how to interpret each of them individually. If you have to formulate your own approaches though, optimization is one of the most powerful tools available for thinking critically about the problem.

Any good numerical computing package will have numerical optimization routines that you can treat as a black box. You input a handle to the function to be minimized and possibly an initial guess as the optimal value. The optimizer will, through some black magic, gradually adjust the initial guess until it is approximately at the optimal value and return it. All numerical optimization routines work this way, by generating a sequence of guesses that (hopefully!) converge to the best solution.

The problem is that this doesn't always work. There are two main ways that optimization fails:

- The sequence of guesses will shoot off infinitely in some direction, rather than converging to an optimal value.
- The guesses will converge on a value, but it is a "local optimum" – a guess that is better than anything nearby it, but worse than another guess at some other location.

It's pretty clear when the first of these happens, and your best bet in that case is to restart it using a different initial guess.

The second problem is much more insidious. The easiest thing to do is try a variety of different initial guesses and see if all the best-performing ones converge to the same place. If so, then this is likely the optimum. Then again, it might not be – you can't really know.

The next section will give you an intuitive idea of what's going on under the hood when a numerical optimization routine runs. It will also explain what's called "convex optimization," which is a large class of optimization problems where algorithms are guaranteed to converge on the correct solution. If you can figure out a way to formulate your optimization problem so that it is convex, you're golden.

23.5 Gradient Descent and Convex Optimization

When trying to understand how an optimization algorithm works, the best picture to have in your mind is to that of walking over a range of hills with your eyes blindfolded. The height of the hills is the value of your objective function, and the x/y coordinates of where you are standing are the components of your guess vector – we are implicitly assuming that $d = 2$ in this example.

There are two questions to answer for every step in the optimization algorithm:

- What direction should you move? Presumably, you want to move downhill, but there are a variety of ways you could decide the exact path.
- How far should you move in that direction?

Pretty much every optimization algorithm boils down to some methodology for answering these two questions. Answer them well, and the hope is that you will find yourself at the bottom of the valley in as few time steps as possible.

Before we go any further, let me define a bit of terminology:

- x will denote a vector of real numbers of length d that we are plugging into the objective function.

- $f()$ will be the objective function.
- The "feasible region" is the set of all x for which all of the $g_i(x)$ are ≤ 0.
- $\nabla f(x)$ will be the gradient of f at x. If you haven't seen gradients before, they're a multivariable calculus thing. $\nabla f(x)$ will be a vector pointing in the direction in which it is increasing most steeply, starting from x; basically, the gradient points directly uphill. The longer the gradient vector, the steeper the increase is.
- As the numerical algorithm progresses, x_0, x_1, ... will be the best guesses at each step.
- x^* will be the solution to the problem.

If the algorithm goes well, then $|x_{n+1} - x_n|$ and $|x^* - x_n|$ will reach 0 as n gets large. If we imagine that x is two dimensional, hopefully the progression of the algorithm will look something like the following:

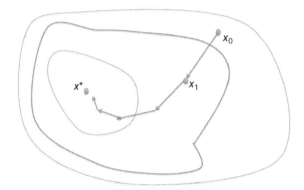

Here are some of the ways that a direction gets picked:

- In some cases, you are able to break out some calculus and derive a closed formula for the gradient of the objective function. If you multiply the gradient by -1, then you have the direction that is pointing the most steeply downhill. This is called "gradient descent," and you might remember it from multivariable calculus.
- In general, you can't get the gradient – all you can do is evaluate the objective function as any point of your choice. In this case, a naive implementation would be to take each dimension in turn, increment your guess for that dimension by a tiny amount, and recalculate the objective function. Then, move in the direction of whichever dimension decreased the objective function the most. This direction will hopefully be relatively close to the gradient.

- If you're really lucky, you can use calculus to derive the "Hessian" of the objective function, which is a matrix giving all of its second partial derivatives. This is also a calculus concept, somewhat more advanced than the gradient. Between the gradient and the Hessian, you can approximate the objective function locally as a paraboloid and move toward the center of the paraboloid.
- The gradient at x_{n+1} is probably pretty close to the gradient at x_n. It is computationally expensive to sample points around x_{n+1} in the hopes of estimating the gradient. But maybe we can keep a running estimate, evaluating $f()$ at only a few points in the neighborhood of x_{n+1} and using that to update our estimated gradient at x_n.
- In a similar way, we can maintain a running estimate of the hessian.

Many algorithms will start with their best attempt at gradient descent, but gradually switch over to using the Hessian as they near the optimal value (the convergence tends to be faster that way).

As to how far we should travel, we typically want to head in the direction we've chosen for as long as it's traveling downhill. Intuitively, we would expect that the distance will be farther if it is currently going downhill more steeply. For this reason, a common technique is to take our estimate of the gradient and use it as the initial step size and try out $x_n + \nabla f(x_n)$. If that lowers $f()$, then try out $x_n + \alpha \nabla f(x_n)$, where α is a parameter greater than 1. Keep multiplying the step size by α for as long as f is decreasing. Conversely, if f was higher $x_n + \nabla f(x_n)$, then divide the step size by alpha until f starts to decrease.

In terms of step size, there are two big problems that we want to avoid. The first is to take small steps and decrease their size so quickly that we take a long time to converge. If we decrease them too quickly, we might even converge to a point that falls short of the minimum as in this picture:

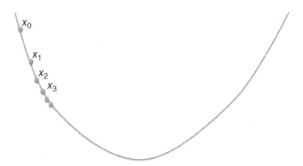

The second problem is if we are regularly overshooting the minimum, as follows:

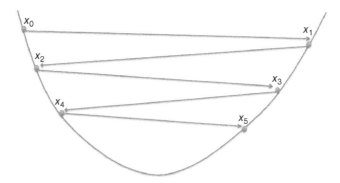

This is a highly inefficient way to reach the minimum. And if we overshoot too far, we can actually increase $f()$!

Even with a solid algorithm that makes all these choices wisely, disaster can still occur. The algorithm will generally converge to *some* locally optimal solution, but nothing I have said here guarantees that it will be *the* best solution. If you start off with x_0 being close to a local optimum, then your algorithm will probably fall into its pit, as in this picture:

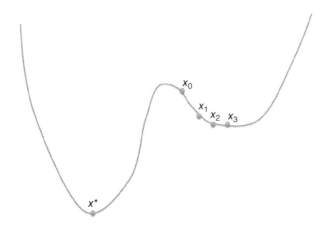

If we want to guarantee that this won't occur, then we will have to look at convex optimization.

23.6 Convex Optimization

An optimization problem is said to be "convex" if it satisfies certain mathematical constraints that guarantee any reasonable numerical algorithm will converge to x^* if it exists. I will give you the technical definition of a convex problem in a second, but intuitively, it means the following:

- The objective function is "bowl-shaped."
- If you follow a straight line from any point in the feasible region to any other point in the feasible region, you will stay entirely within the feasible region.

The first constraint means that the function has no local minima expect for the single global minimum, sitting in the middle of the "bowl." You can't fall into a trap like in the previous figure.

The second constraint means that there is a line from x_n to x^* that lies within the feasible region. If that weren't the case, then you could have a situation like the following:

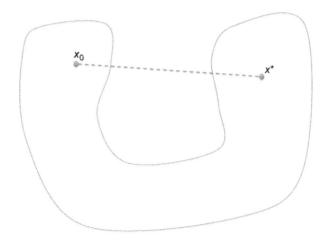

where I've drawn the border of the feasible region in a solid line. Gradient descent will probably bring our algorithm up against the border of the feasible region, and it will stop there. The function $f(x)$ does not have a local minimum, but it does if we limit ourselves to this nonconvex feasible region.

If f is bowl-shaped, then it is called "convex." If the feasible region meets our criteria, then it is also called "convex." Yes, people overuse this word.

A region of space is convex if for any two points x and y in it, the line segment from x to y lies in the region as well. Think of it this way: if you wrapped Saran Wrap tightly over the region, the Saran Wrap would touch the region at every point.

The function f is said to be convex if, for any points x and y in space and any number α between 0 and 1,

$$f\big(\alpha x + (1-\alpha)y\big) \le \alpha f(x) + (1-\alpha)f(y)$$

Convexity is a very, very restrictive condition. If you write out a random function, it is very unlikely to be convex. A lot of people spend a lot of time reformulating nonconvex problems into equivalent ones that are convex or proving that certain classes of problems are convex. Theoretically, there's almost no middle ground: if the problem is convex, then any half-way decent optimization algorithm will converge to the right answer (the good algorithms will just do it faster). If it is not convex, then you have no guarantees.

That doesn't mean that you should give up hope though if your problem isn't convex! Much of the work in machine learning involves problems that aren't convex (or at least, nobody has proven that they are), and a lot of excellent work has been done. It's just that in place of rock-solid theorems, we must make do with rules of thumb and empirical findings. For some classes of problems, optimization algorithms work shockingly well for something that is not convex, and it's an open question why.

23.7 Stochastic Gradient Descent

In many optimization problems, the objective function is a sum of many much simpler functions. For example, in MLE, we often try to optimize

$$\log\big(P(X\mid\theta)\big) = \sum_{i=1}^{N}\log\big(P(X_i\mid\theta)\big)$$

where θ is the collection of parameters for our model (of which there are probably only a few), and N is the (often very large) number of points in our training dataset. Calculating the gradient of $\log\big(P(X\mid\theta)\big)$ is probably computationally prohibitive.

The idea of stochastic gradient descent (SGD) is that we can approximate the gradient by picking a single value for i and then calculating the gradient of $\log\big(P(X_i\mid\theta)\big)$. Each step in the algorithm picks a new i, either by selecting randomly or sweeping through the entire dataset.

A variation of SGD is "mini batch" SGD. In this, we take a selection of more than one data point and calculate the gradient based on them. The idea is to pick enough points that we get a better idea of the actual gradient at a point, but few enough that calculating the gradient is still computationally feasible. This tends to converge faster than vanilla SGD.

23.8 Further Reading

1 Boyd, S & Vandenberghe, L, *Convex Optimization*, 2004, Cambridge University Press, Cambridge, UK.
2 Nocedal, J & Wright, S, *Numerical Optimization*, 2nd edn, 2006, Springer, New York, NY.

23.9   Glossary

Convex optimization An optimization problem where the feasible region and the objective function are both convex. This guarantees that most optimization algorithms will converge to the global optimum.

Convex function Intuitively, this is a function that is "bowl-shaped."

Convex region A region of d-dimensional space such that if two points x and y are within the region, then the line segment between x and y is also in the region.

Global optimum The location in a feasible region that minimizes the objective function.

Gradient Say, you have a function f that takes in a d-dimensional vector x and outputs a number. Then, the gradient of f at x is a vector pointing in the direction in which f is increasing most steeply. The longer the gradient is, the more steeply f increases in that direction. The gradient of a function can often be calculated using the tools of multivariable calculus.

Gradient descent An optimization algorithm that attempts to calculate the gradient of the objective function at a guessed point x, then reduce the objective function by traveling along that gradient to a new point, and repeat. Generally, this process will converge to a locally optimal solution.

Optimization A field in numerical computing that focuses on finding the input vector that minimizes some objective function, possibly subject to certain constraints on the input.

Objective function A function that takes in a vector and outputs a real number. The goal in optimization is to find the input that minimizes the objective function.

Local optimum A point x in the feasible region where the objective function is lower than (or equal to) its value at any point in a certain radius of x. It may not, however, be the best point in the entire feasible region.

Maximum likelihood estimation Fitting a probability distribution to real-world data by setting the parameters of the distribution (such as the mean and standard deviation, in the case of fitting a Gaussian) so as to maximize the probability of observing the data we got.

MLE Maximum likelihood estimation

Stochastic gradient descent An optimization method where we estimate the gradient by taking a random selection of data points and calculating the gradient only based on those points.

24

Advanced Classifiers

This chapter will dive deeper into some more cutting-edge machine learning algorithms. In the first part of this book, I had a chapter on machine learning classification, so why didn't I put this material there?

My rule was that the classifiers in the earlier chapter could be used pretty much as black boxes. Yes, it was certainly possible to dissect them, analyze why they worked well or poorly, use them to extract fancier features, and so on. But you *could* just use them out of the box, and they would probably work pretty well.

The classifiers in this chapter are something of a dark art and really designed more for machine learning specialists than normal data scientists. They have a lot of internal structure, which must be planned out by the user, there are many different parameters to tune (arbitrarily many, depending on how complex the structure is), and there is no good precanned answer about how to make those decisions. You will have to think critically about the problems you plan on solving, design a classifier that is appropriate for those problems, and then experiment with different parameters and layouts to find what works best. And if you screw it up somehow, your classifier is liable to perform horribly.

In exchange for this extra work, you get to use the most powerful types of classifiers in the world today. This is the kind of stuff going on the heart of Google and Microsoft. It is used for hard problems such as identifying the people in a random Facebook photo, Siri figuring out what you want based on a garbled command spoken into your phone, and deciding whether an image is safe enough to show random people on an image search. Soon, algorithms such as this will probably be driving your car.

The two types of classifiers I will discuss here are deep learning and Bayesian networks. Deep learning is a broad class of next-generation neural networks, which were made possible by several advancements in the numerical techniques required for training them. The second class of models is Bayesian networks.

The Data Science Handbook, First Edition. Field Cady.
© 2017 John Wiley & Sons, Inc. Published 2017 by John Wiley & Sons, Inc.

Since these are deep topics, which data scientists rarely need anyway, the coverage in this chapter will be relatively cursory. I will focus mostly on example code that illustrates the key ideas and how to play around with them.

Finally, I should give you a personal qualifier. I want to make it clear that I am not an expert in these techniques. Data science is a big field, and the subjects in this chapter are things that I haven't had a lot of occasion to use when solving real problems. But I do know enough to give you the key ideas, show you how to write some simple applications, and point you in the right direction if you want to learn more.

24.1   A Note on Libraries

I've tried hard in this book to give example code using libraries that were relatively standardized. They work well, they are widely used and trusted, and I expect that people will still be using them 5 years from now.

There are no such libraries for the topics I'm discussing in this chapter. There are multiple competing ones, and it remains to be seen which software packages will come out on top. I will still show you example code, written with the libraries that I myself use, but you should be aware that these libraries are still in their infancy. The documentation is bad, there are bugs, and the APIs are likely to change in the future. And there is a good chance that these libraries will become obsolete as momentum shifts over to one of their competitors.

The leading libraries for deep learning are Theano and TensorFlow. However, they are extremely low-level libraries designed for people with a deep knowledge of the math that underlies deep learning. Instead, I will use Keras, which is a more user-friendly (but not *too* user-friendly – this is still deep learning) wrapper around them. Keras lets you focus on how to structure your deep network, rather than the numerical details of how it gets trained.

For Bayesian networks, I use PyMC. This library was originally created to help people use a numerical technique called Markov chain Monte Carlo (MCMC) simulation. MCMC is useful for a variety of applications, and one of them is Bayesian networks. This is the application that really caught on for PyMC, and today you can use the library with almost no awareness of the MCMC algorithms running in the background.

24.2   Basic Deep Learning

Recall from the previous chapter that the simplest neural network is the perceptron. A perceptron is a network of "neurons," each of which takes in multiple inputs and produces a single output. An example is shown in the following figure:

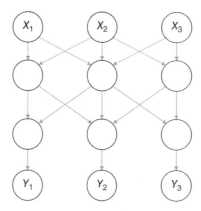

The labeled nodes correspond to either the input variables to a classification or a range of output variables. The other nodes are neurons. The neurons in the first layer take all of the raw features as inputs. Their outputs are fed as inputs to the second layer and so on. Ultimately, the outputs of the final layer constitute the output of your program. All layers of neurons before the last one are called "hidden" layers.

A perceptron has only a single hidden layer. The simplest type of deep learning is what are called "deep networks" – neural nets with multiple hidden layers.

Rather than getting into any more theory for now, let's show the code for building the simple neural net in the picture:

```
from keras.models import Sequential
from keras.layers.core import Dense, Activation
model = Sequential([
    Dense(3, input_dim=3, activation='sigmoid'),
    Dense(3, activation='sigmoid')
])
model.compile(
    loss='categorical_crossentropy',
    optimizer='adadelta')
```

At this point, the model will have the normal train() and test() methods that we are familiar with from machine learning.

Let's step through what's happening:

- The simplest type of neural network in Keras is the Sequential model, which means that there are layers of neurons stacked on each other with each layer feeding into the next.

- The first layer we see is a Dense. The input_dim = 3 means that there are three nodes in the previous layer. Each of those nodes will have its output directed to every node in the current layer. There will also be three nodes in the current layer.
- The activation parameter says how each node should condense all of its inputs into a single output. In this case, we use the usual sigmoid activation function, though there are others. This completes our hidden layer of neurons.
- There is another dense layer, routing all the hidden outputs to each of three nodes. Note that we don't need to specify the input_dim here, since it is equal to the dimensionality of the output of the previous layer.
- Finally, we compile the model into a low-level implementation in Theano or TensorFlow. This is not the same thing as training it – we are just configuring the number-crunching machinery so that it can be trained. The "loss" parameter says what type of objective function should be minimized when it comes time to train the model. The "optimizer" specifies an algorithm for doing the actual training.

Here is a more complete code example, which constructs a neural net with 50 neurons in a single hidden layer and uses it to classify the data points in the Iris dataset. Note the preprocessing step where I turn the flower's species into a three-dimensional vector, with the dimension corresponding to the correct species set to 1.

```
import sklearn.datasets
from keras.models import Sequential
from keras.layers.core import Dense, Activation
import pandas as pd
from sklearn.cross_validation import train_test_split
ds = sklearn.datasets.load_iris()
X = ds['data']
Y = pd.get_dummies(ds['target']).as_matrix()
X_train, X_test, Y_train, Y_test = \
    train_test_split(X, Y, test_size=.2)
model = Sequential([
    Dense(50, input_dim=4, activation='sigmoid'),
    Dense(3, activation='softmax')
])
model.compile(
  loss='categorical_crossentropy',
  optimizer='adadelta')
model.fit(X_train, Y_train, nb_epoch=5)
proba = model.predict_proba(X_test, batch_size=32)
```

```
pred = pd.Series(proba.flatten())
true = pd.Series(Y_test.flatten())
print "Correlation:", pred.corr(true)
```

To give you a little idea of how finicky neural networks can be, it took me a while to put this example code together. If I had 10 neurons in the hidden layers rather than 50, for example, the performance would vacillate randomly between different times I trained/tested. Sometimes, the correlations would be up to 0.6. Other times, they would be very close to 0 or even negative a few times. I found empirically that the having 50 nodes makes the performance much more stable around 0.6.

24.3   Convolutional Neural Networks

One of the most important extensions of neural networks, which is especially important when classifying images or sounds, is the convolutional network. In software, they take the form of having a "convolutional" layer.

Convolutional networks are partly inspired by human neurology. In your brain, in the earliest stages of processing images, there are neurons that specialize in very primitive patterns in very specific parts of the visual field. There might be one that activates whenever there is the shape of a line sloping up in a particular part of your upper-right field of view. In fact, neurons corresponding to nearby regions of the visual field tend to be nearby each other in your cerebral cortex. In this way, their activation patterns take what you're looking at and paint a distorted version of it right onto the surface of your brain. Later parts of your visual system will do the same thing to the previous layers. If you'll permit me to grossly oversimplify, stage 1 might have neurons that detect straight lines at various slopes, in different parts of the visual field. Stage 2 will have neurons that combine adjacent neurons from stage 1 and fire if those sloped lines fit together to form a square and so on. At the highest levels, you will have neurons that fire in response to something as specific as a picture of Homer Simpson.

A convolutional layer in a neural network has several "filters," each of which looks for a particular pattern in the image. Each filter has a "kernel": a small matrix of numbers, which, when seen as an image, resemble the pattern that this filter is detecting. To detect the pattern at various parts of the image, we slide the kernel all over the image and take the dot product of the kernel with the pixels that it is overlapping. Typically, we will slide it up/down by a fixed number of pixels and left/right by a fixed (possibly different) number of pixels. This process is called a "convolution" between the original image and the kernel.

A simple version of convolution is shown in this image, where we move the kernel by three pixels every time we slide it:

Kernel

0	1	0	0
0	1	0	0
0	0	1	0
0	0	0	1

Image

0	2	0	0	0	0	0
0	0	0	0	0	0	0
0	0	0	0	0	3	0
0	1	0	0	0	0	0
0	0	0	0	0	0	0
0	0	0	3	0	0	0
0	0	0	0	0	0	0
0	0	0	0	0	0	0

Convolution

2	3
1	0

The values in the convolved image now represent the degree to which the image, at that point, looked like the kernel. In a neural network, you can have one or more convolutional layers feeding into each other, and each of them can have several different filters.

The image can also have more than one number associated with each pixel. For example, in a typical colored image, there will be three values giving the amounts of red, green, and blue at that point. In this case, the kernels will be three-dimensional arrays.

I have also presented this in the context of convolutions over images. While that's the most famous application, the mathematics of convolution works equally well for a one-dimensional signal such as a time series. In areas such as this, convolutional nets are useful for areas such as speech recognition.

24.4   Different Types of Layers. What the Heck Is a Tensor?

The name of the software TensorFlow speaks to the mathematical underpinnings of neural nets, and it's important to have at least some idea of what's going on there. The word "tensor" has a number of technical definitions, but colloquially, it is often used to describe multidimensional arrays of numbers that we do linear operations on. Understanding this perspective is important for actually constructing complicated neural nets, so I would like to review it.

The input to a basic neural net is a one-dimensional array, and its output is also a one-dimensional array. The input to a two-dimensional convolutional network is a two-dimensional array, and its output is one-dimensional. In general, every layer of a network takes a tensor as input and produces a tensor as its output.

Most of what goes on in a neural net consists of linear operations on these tensors. Take a plain-vanilla layer with a sigmoid activation function. It operates in the following way:

- It takes in inputs, which we can think of as a tensor.
- Every node takes a weighted combination of its inputs. Alternatively, it takes a weighted combination of *all* the inputs, except that many of the coefficients are 0.
- If there are d inputs and n nodes, and each node takes a linear combination of the inputs, this is really just multiplying an n-by-d matrix by the input vector.
- Finally, we apply the sigmoid function to every element in the weighted-sum vector. This is the only part that's not linear.

The layer of the network is completely characterized by the matrix of its weight coefficients. In terms of implementation, most of what you're doing boils down to linear algebra.

A convolutional layer in a neural network is entirely linear. Conceptually, it takes in a two-dimensional tensor, multiplies it by a four-dimensional tensor, and has the output that is another two-dimensional tensor.

Note that the input to a convolutional net is 2D, but the final output will typically be 1D. So, there will need to be some layer where we switch from two to one dimension. Typically, this layer is also completely linear, a flattening layer that takes in a two-dimensional tensor of dimensions d-by-n and outputs a 1D tensor of length $d*n$ containing all the same values. This is just multiplying the input tensor by a three-dimensional tensor, all of whose values are either 0 or 1.

24.5 Example: The MNIST Handwriting Dataset

The standard dataset for learning about convolutional nets is called MNIST, which contains 70,000 images of handwritten digits, each one 28 pixels by 28 pixels. This section will walk through running a convolutional network on the MNIST data.

A gotcha that I should clarify upfront is that in Keras, 2-d convolutional neural nets expect an input array with four dimensions:

- The first dimension is for which data point you're looking at.
- The second dimension is for the "channel." In my code, there is only one channel, since my images have only one floating number for each pixel. In color images, you are likely to have three channels, for red, green, and blue. In general, you could have more.
- The third and fourth dimensions are the number of rows and columns in the input images.

This example will use all the layer types we have discussed so far and once other: dropout. A dropout layer is a heavy-handed way to reduce overfitting in neural nets. A dropout layer takes in a parameter p, denoting the probability of zeroing-out a value. During each training iteration, the dropout layer will take in a tensor, set a random fraction p of its values to 0, and then pass the array on the next layer for training.

The following code will fetch the MNIST dataset and train a convolutional neural network on it. Be aware that you must be connected to the Internet for this code to run, and it will take quite a while to do so, since it fetches the MNIST dataset at runtime.

```
import theano
import statsmodels.api as sm
import sklearn.datasets as datasets
import keras
from keras.models import Sequential
from keras.layers.core import Dense, Activation,
Dropout, Flatten
import pandas as pd
from matplotlib import pyplot as plt
import sklearn.datasets
from keras.layers.convolutional import Convolution2D,
MaxPooling2D
from sklearn.cross_validation import train_test_split

from sklearn.decomposition import PCA
from sklearn.cluster import KMeans
from sklearn.metrics import silhouette_score,
adjusted_rand_score
from sklearn import metrics
from sklearn.cross_validation import train_test_split
from sklearn import datasets
import sklearn
from sklearn.datasets import fetch_mldata
```

```python
dataDict = datasets.fetch_mldata('MNIST Original')
X = dataDict['data']
Y = dataDict['target']
X_train, X_test, Y_train, Y_test = \
    train_test_split(X, Y, test_size=.1)
X_train = X_train.reshape((63000,1,28,28))
X_test = X_test.reshape((7000,1,28,28))
Y_train = pd.get_dummies(Y_train).as_matrix()

# Convolution layers expect a 4-D input so we reshape
our 2-D input
nb_samples = X_train.shape[0]
nb_classes = Y_train.shape[1]

# We set some hyperparameters
BATCH_SIZE = 16
KERNEL_WIDTH = 5
KERNEL_HEIGHT = 5
STRIDE = 1
N_FILTERS = 10

# We fit the model
model = Sequential()
model.add(Convolution2D(
    nb_filter=N_FILTERS,
    input_shape=(1,28,28),
    nb_row=KERNEL_HEIGHT,
    nb_col=KERNEL_WIDTH,
    subsample=(STRIDE, STRIDE))
)
model.add(Activation('relu'))
model.add(MaxPooling2D(pool_size=(5,5)))
model.add(Dropout(0.5))
model.add(Flatten())
model.add(Dense(nb_classes))
model.add(Activation('softmax'))
model.compile(loss='categorical_crossentropy',
optimizer='adadelta')

print 'fitting model'
model.fit(X_train, Y_train, nb_epoch=10)
probs = model.predict_proba(X_test)
preds = model.predict(X_test)
pred_classes = model.predict_classes(X_test)
true_classes = Y_test
(pred_classes==true_classes).sum()
```

24.6   Recurrent Neural Networks

One of the big things about neural nets so far in that, similarly to normal machine learning models, they handle the points that they classify independently. If they are fed a series of points, they will classify each one with no reference to any of the others. Essentially, they have no memory after they're done processing a point.

Recurrent neural networks (RNNs) have a primitive type of memory, in the form of "recurrent" layers. A recurrent layer takes in two kinds of input: the output of the preceding layer and the output of the same recurrent layer from the last point it processed. That is, the output of a recurrent layer when classifying a point will be passed back into this layer when classifying the next point.

Recurrent nets are thus ideal for dealing with time series data. Ascribing specific meaning to the output of a recurrent layer is very difficult, but just to give you an idea: the output could encode a prediction about what the next point will be, given the current point. It could also encode a long-term memory of some point that was seen a long time ago.

There are many types of recurrent layers available, which keep a variety of different types of memory between classifications. The ones you are most likely to see are as follows:

- Basic recurrent nets: This will just recall the last output of the layer and feed it back as input during the next classification.
- LSTMs. This stands for "long short-term memory," and it is very good at clearly remembering select events from very far in the past. This is in contrast to basic RNNs, for which the memory of an event decays over time.

The following example code uses a built-in time series dataset as an example for training and testing a recurrent neural net in Keras. The training data is a tensor of shape (num_sequences, timestamps_per_sequence, num_dimensions). In this case, each vector is only of dimension 1, but there are 11 measurements in each sequence.

```
import sklearn.datasets
from keras.models import Sequential
from keras.layers.core import Dense, Activation
from keras.layers.recurrent import LSTM, GRU

import statsmodels as sm

df = sm.datasets.elnino.load_pandas().data
X = df.as_matrix()[:,1:-1]
X = (X-X.min()) / (X.max()-X.min())
Y = df.as_matrix()[:,-1].reshape(61)
Y = (Y-Y.min()) / (Y.max()-Y.min())
```

```
X_train, X_test, Y_train, Y_test = (
  train_test_split(X, Y, test_size=0.1)
model = Sequential()
model.add(GRU(20, input_shape=(11,1)))
model.add(Dense(1, activation='sigmoid'))
model.compile(loss='mean_squared_error',
optimizer='adadelta')
model.fit(X_train.reshape((54,11,1)),
  Y_train, nb_epoch=5)
proba = model.predict_proba(X_test.reshape((7,11,1)),
batch_size=32)
pred = pd.Series(proba.flatten())
true = pd.Series(Y_test.flatten())
print "Corr. of preds and truth:", pred.corr(true)
```

24.7 Bayesian Networks

Let's briefly recap about Bayesian statistics from earlier in the book. We have a random variable $D = (D_1, D_2, ..., D_d)$ that is considered to represent our data, and we have a random variable Y that represents whatever underlying thing we are trying to infer. Intuitively, we want to get our hands on $P(Y|X)$, the probability of Y having a certain value, given the data X that we observe. The key equation of Bayesian statistics is

$$P(Y|D) = \frac{P(D|Y)P(Y)}{P(D)}$$

Taking the terms on the right, we have the following:

- $P(Y)$ is called the "prior," and it is our initial confidence in the different values Y could have. This is the philosophical controversial part of Bayesian stats, because $P(Y)$ is really just a measure of our subjective confidence, but we treat it as a mathematical probability.
- $P(D|Y)$ is where the magic happens. This is the probability of the data we got, assuming an underlying value of Y. In practice, this is a LOT easier to think about and to model than $P(Y|D)$. Most of the work in doing Bayesian analysis involves the question of how we model $P(D|Y)$ and fit the model to our data.
- $P(D)$ is the probability of seeing our data D. To get this probability, you must average over the different possible values of Y, which can be quite complicated. However, in practice, you almost never need to actually calculate $P(D)$, since it's just a normalization term so that all of the $P(Y|D)$ add up to 1.

Usually, you're only interested in the relative probabilities of the different values of Y.

The term $P(D|Y)$ is the big complicated one, since in general, the various D_i could have complicated dependency structures between them. The naive Bayes assumption is that

$$P(D|Y) = P(D_1|Y) * P(D_2|Y) * \cdots * P(D_d|Y)$$

that is, the D_i are all conditionally independent of each other, if we know Y. In general though, this is not true, and we represent the dependencies between the variables with a "Bayesian network" such as the following:

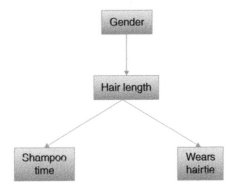

This network says that knowing somebody's gender tells you something about how long their hair is likely to be, and hair length in turn can be used to predict whether somebody wears a hair tie. It may be that a woman is more likely to wear a hair tie than a male, but that's only because women are more likely to have long hair. Among the people who have long hair, there is no relationship between gender and hair ties, and similarly among those who have short hair.

When you are using Bayesian networks, generally it is up to you as the user to pick the topology of the network, along with some probabilistic model of how each variable depends on the ones that influence it. Training data is used to fit the actual parameters of the dependencies.

Network topology is a place where expert knowledge or intuition can be inserted into the model. In this sense, Bayesian networks are almost the opposite of neural nets; they lend themselves to the introduction of intuition and deep understanding, whereas neural nets are nearly impossible to make sense of.

Similar to other machine learning models, there are two ways you might want to use Bayesian network models: either you can use them to make predictions or you can dissect the trained models to get insights about whatever it is that you're studying.

24.8 Training and Prediction

When we discussed naive Bayes classifiers in a previous chapter, we had a prior confidence $P(Y)$ over all classifications, but then we had precise models $P(D|Y)$ of what the data would look like conditioned on different classes. The model parameters that characterize the $P(D|Y)$ are known precisely. Typically, those parameters are learned using maximum likelihood estimators on the training data, which opens the door to overfitting issues.

In general, with Bayesian networks though, the model parameters themselves have a confidence distribution over them. Before training, we have priors over what those parameters might be, and we refine our confidence in light of the training data. When it comes time to classify points in the future, we must average our predictions over all possible model configurations, weighting them by our confidence. This means that the probability will be proportional to a very complicated integral:

$$P(Y|D) \propto P(Y)\int_{\theta} \text{Confidence}(\theta)P(D|Y,\theta)\mathrm{d}\theta$$

where Confidence(θ) bakes in both our priors over θ and our training data. This is a very complicated (and sometimes impossible) integral to perform, and practitioners of Bayesian networks often use numerical approximations instead of just calculating it directly.

24.9 Markov Chain Monte Carlo

PyMC is a Python library for fitting Bayesian networks to real data. Unfortunately, it is not a classifier in the normal sense. If you want to use the fitted model to actually make predictions, then you will have to do it on your own. However, PyMC will do the heavy lifting of fitting a potentially extremely complicated model, and that's where the bulk of the work is.

Let's call Θ the correctly trained distribution over all values of θ, taking into account both our priors and the data we train on. The probability density for Θ at θ will be Confidence(θ) from the integral in the previous section. Instead of giving us the function Confidence(θ) directly, which is computationally intractable, PyMC gives us random samples from the distribution Θ. You can then estimate the integral by simply summing over all of the samples.

I don't want to get too far into the weeds here. Without giving you too much theory, PyMC fits Θ using a technique called "Markov chain Monte Carlo" (MCMC). Running an MCMC simulation will give us a sequence of guesses at θ that will, over many time stamps, be sampled from the trained distribution Θ. Let's call the ith guess θ_i. That means, for example, that the average of many MCMC θ_i is guaranteed to converge to the actual average value of Θ.

What MCMC does *not* guarantee, however, is that subsequent guesses are independent samples from Θ. Quite to the contrary, each guess is correlated with the ones before and after it. This means that θ_n will be correlated with θ_{n+1}, which will in turn be correlated with θ_{n+2}. Then θ_n will also be correlated with θ_{n+2}, just not as strongly as it is with θ_{n+2}. If the correlations decay quickly, then the θ_i will be nearly independent of each other. But if they are very highly correlated, then it might take many, many samples before we get a decent approximation of Θ.

If you think of the probability distribution of Θ as a landscape, imagine the MCMC process as taking a random walk over the landscape that is guaranteed to spend more time in the likelier regions. You have a mathematical guarantee that, in the limit of walking for an infinite amount of time, the amount of time you spend in any area will be proportional to the probability of Θ in that area. But in practice, depending on where you start and how large your steps are, it could take you arbitrarily long to actually get to a particular high-probability area.

This autocorrelation of the θ_i is the biggest problem with MCMC models. There are internal mechanisms of a MCMC model that you can tune, which might reduce the autocorrelation. But doing this well is tricky, and there's not much that has any guarantees of working.

Now as with all theoretical boogey men, this is often not a problem in practice. In many cases, your first few θ_i will be grossly wrong because you started at an unlikely location, but soon they converge reliably to a best-fit value that they rarely deviate far from.

Starting from an unlikely situation though is a problem, since it will skew any statistics that you care to calculate. So, what is often done is that you pick a suitably large number N and then discard θ_1 to θ_N, the so-called burn in. After that the distribution is assumed to be "stable enough."

24.10   PyMC Example

Scikit-learn has a toy dataset built in that gives housing statistics for several towns in the Boston area. I will pull out several of those statistics and show you PyMC code that fits the following Bayesian network:

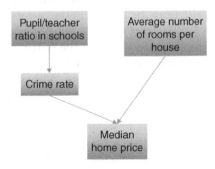

The idea here is that a poor educational system leads to higher crime rates, which in turn will reduce home prices. In reality, bad schools will also impact home prices directly, but I'm ignoring that for this model.

I will make the following assumptions about the variables and their relationships:

- The pupil/teacher ratio is an exponential distribution, whose mean I know from the data.
- The crime for a city will also be exponentially distributed. However, its mean will be some constant A times the pupil/teacher ratio. We will need to fit A from the data.
- The average number of rooms for a town will be normally distributed, with the mean and standard deviation present in the data.
- The median home price for a town will be normally distributed, with a mean of B*{avg. number of rooms}*(C-{crime rate}). Here B and C are unknown parameters that must be fitted.
- A, B, and C will all have exponentially distributed priors with mean 1.

In order to use PyMC, you will need to declare the model in its entirety in a Python file, which we will call mymodel.py. Put the following code into mymodel.py:

```
import pandas as pd
import sklearn.datasets as ds
import pymc
# Make Pandas dataframe
bs = ds.load_boston()
df = pd.DataFrame(bs.data, columns=bs.feature_names)
df['MEDV'] = bs.target
# Unknown parameters are A, B and C
A = pymc.Exponential('A', beta=1)
B = pymc.Exponential('B', beta=1)
C = pymc.Exponential('C', beta=1)
ptratio = pymc.Exponential(
    'ptratio', beta=df.PTRATIO.mean(),
    observed=True, value=df.PTRATIO)
crim = pymc.Exponential('crim',
    beta=A*ptratio, observed=True, value=df.CRIM)
rm = pymc.Normal('rm', mu=df.RM.mean(),
    tau=1/(df.RM.std()**2), value=df.RM, observed=True)
medv = pymc.Normal('medv', mu=B*rm*(C-crim),
    value=df.MEDV, observed=True)
```

Note a few things about this code:

- Whenever we declare a variable, we specify its family and the parameters that characterize it.
- You can make the parameters for a variable take on a fixed value. Or, you can make them take on a value that is defined by other variables.
- If the data were actually observed, you can say observed = True and pass in the observed values.

Once this file is in place, you can train it this way:

```
import pymc
import mymodel
import matplotlib.pyplot as plt
S = pymc.MCMC(mymodel)
S.sample(iter = 40000, burn = 30000)
pymc.Matplot.plot(S)
plt.show()
```

This says to discard the first 30 thousand simulations and then calculate statistics based on the next 40 thousand. It will produce plots such as the following one, which shows the evolution of B over the course of the simulation:

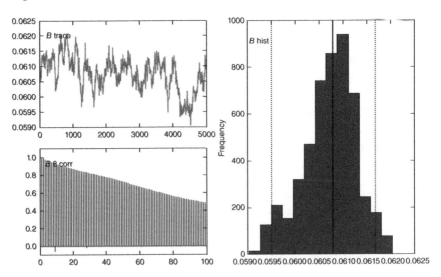

We can see on the right, a histogram of all *B* values, which are centered pretty tightly around 0.0605. On the upper left is the evolution of *B* over the course of the data we gathered. From the plot it seems that the simulation has equilibrated, and we can probably make use of the numbers. Technically speaking, there is a risk that there is still some large area of Θ that our simulation hasn't wandered into yet, and running it a little longer would change the dynamics dramatically. But looking at the graphs that sure seems unlikely.

24.11 Further Reading

Koller, D & Friendman, N, *Probabilistic Graphical Models: Principles and Techniques*, 2009, The MIT Press, Cambridge, MA.
Goodfellow, I, Bengio, Y & Courville, A, *Deep Learning*, 2016, MIT Press, http://www.deeplearningbook.org/.

24.12 Glossary

Bayesian network An arrangement of random variables into a directed graph structure, where each graph is conditionally independent of all nodes above it except its immediate parents.
Convolutional neural network A neural net where one of the layers takes the convolution of the input with a kernel. Typically, a kernel will capture a specific, relatively low-level pattern in the data. The output of the convolutional layer will be an indication of how much that pattern is present at each part of the input data.
Deep neural net A neural net with an unusually high number of layers.
Deep learning A broad term for a wide range of developments that have happened recently in neural networks.
Markov chain Monte Carlo A simulation technique that is useful for fitting the parameters of Bayesian network models.
Neural network A class of machine learning models inspired by the wiring of neurons in biological brains.
Recurrent neural network A neural network where the output of a layer can be fed back into that layer.
Tensor An array of floating-point numbers. There can be arbitrarily many dimensions to the array.

25

Stochastic Modeling

Stochastic modeling refers to a collection of advanced probability tools for studying not a single random variable, but instead, a *process* that happens randomly over time. This could be the movement of a stock price over time, visitors arriving at a web page, or a machine moving between internal states over the course of its operation.

You're not locked in to using time either; anything that is sequential can be studied. This includes which words follow which others in a piece of text, changes from one generation to the next in a line of animals, and how temperature varies across a landscape. The first place I ever used stochastic analysis was studying the sequence of nucleotides in DNA.

This chapter will give you an overview of several of the main probability models, starting with the most important one: the Markov chain. I will discuss how they are related to each other, what situations they describe, and what kinds of problems you can solve with them.

25.1 Markov Chains

By far the most important stochastic process to understand is the Markov chain. A Markov chain is a sequence of random variables X_1, X_2, ... that are interpreted as the state of a system at sequential points in time. For now, assume that the X_i are discrete RVs that can take on only a finite number of values.

Each of the X_i has the same set of states that it can take on. The definitive feature of a Markov chain is that the distribution of X_{i+1} can be influenced by X_i, but it is independent of all the previous RVs if you know X_i. That is,

$$P(X_{i+1} = x \mid X_1, X_2, ..., X_i) = P(X_{i+1} = x \mid X_i)$$

The cartoonish way I like to think of this is that you have a collection of lily pads and a frog jumping between them randomly. At any point in time, the frog

The Data Science Handbook, First Edition. Field Cady.
© 2017 John Wiley & Sons, Inc. Published 2017 by John Wiley & Sons, Inc.

knows which lily pad it is on, and this knowledge determines how likely it is to jump to every other lily pad at the next step (or it could just stay on the current lily pad). But the frog has amnesia; it has no memory of what lily pad it was on at any previous step or how long it has been hopping around. So, the probability distribution of its next hop is only a function of where it is now.

If there are k different lily pads, we arrange these transition probabilities into a k-by-k matrix. The ith row corresponds to being on the ith lily pad. The jth entry in that row will be the probability of jumping to the jth lily pad, given that we are currently on the ith one. This means that every entry in the transition matrix must be greater than or equal to 0, and every row sums to 1.0. Knowing the transition matrix and the initial probability distribution of X_1 completely characterizes the Markov chain.

It is common to draw our Markov chain, especially ones where k is small, in diagrams with arrows pointing between the states. Here is one for a particular Markov chain I concocted describing the weather:

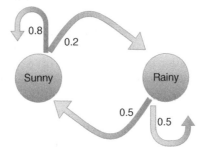

In this model, a sunny day has an 80% probability of being followed by another sunny day, and the day after it rains will be 50/50 rain or shine. The transition matrix for the Markov chain looks like the following:

	Sunny	Rainy
Sunny	0.8	0.2
Rainy	0.5	0.5

If there are k states in the system and we are in the ith state, then we can express that as a k-by-1 row vector that is equal to 1.0 at i and 0 everywhere else. Call the state vector p. The vector p can also have several nonzero components, just so longer as they are all nonnegative and they sum to 1. In this case, p represents a probability distribution over the possible states. If we want to know the distribution at the next time step, we simply multiply p by the transition matrix T:

$$p_{i+1} = p_i T$$

and in general

$$p_{i+m} = p_i T^m$$

The fact that X_{i+1} is only influenced by X_i is called the "Markov assumption." It is the key to making Markov chains tractable, both mathematically and computationally. The Markov assumption is implicit in the state transition diagram, because each state has no indication of how you got there. It only shows where you might be going next.

In many applications, the Markov assumption does not hold. For instance, it is common to crudely model natural language by having the X_i represent words in a piece of text. If there is a word, this is missing or ambiguous, a Markov chain can be used to find what the word most likely is (more on that later). The problem is that a Markov chain that only incorporates one word of context is woefully inadequate.

The typical solution is to have X_{i+1} depend not only on X_i, but also on several of the values before that back to X_{i-m-1}. This is called a Markov chain of order m. Strictly speaking though, an order-m Markov chain is equivalent to a beefed-up Markov chain of order 1. If we define a new set of random variables

$$Y_1 = (X_1, X_2, \ldots, X_m)$$
$$Y_2 = (X_2, X_3, \ldots, X_{m+1})$$
$$Y_3 = (X_3, X_4, \ldots, X_{m+2})$$

then the Y_i will behave as an order-m Markov chain. It will just be one with k^m states, and all of the numbers in the state transitions that would overwrite the previous X_i have probability 0. In industrial scale, Markov chains for natural language, something such as $m = 6$ is common.

There are a lot of words in the English language. That number raised to the sixth power is astronomically large. This should give you an idea of some of the practical problems in working with large Markov chains. When the number of states gets very large, it is impossible to get enough data to fit the transition matrix, and it becomes computationally intractable to explore them all. In these situations, there are a variety of heuristics and performance optimizations that get utilized.

25.2 Two Kinds of Markov Chain, Two Kinds of Questions

Most practical Markov chains come in two types: "irreducible aperiodic" and "absorbing." They have different mathematical properties and are used for modeling very different types of systems.

An irreducible aperiodic Markov chain has two key properties:

- It is possible to get from any state to any other state in a finite number of time steps.
- No state is "periodic." That is, you don't have a weird situation where you can only ever return to a particular state after an odd number of steps or something similar. If you look far enough into the future, it is possible to be in any state at any time.

These properties guarantee that no matter what state the Markov chain starts in it will, in the very long run, have a "steady state" behavior where it could be in any state. The long-term distribution of how likely it is to be in each state is also called the "equilibrium distribution." An irreducible aperiodic Markov chain is guaranteed to have a unique equilibrium distribution. The distribution is a k-dimensional vector of probabilities, where the ith component gives the probability of being in the ith state. Typically, this vector is referred to as π.

The weather diagram I drew was irreducible and aperiodic: no matter what the weather is today, the probability of it being rainy in 1000 days is functionally just the fraction of all days when it is sunny. Natural language is another example. No matter what word we are looking at now, in a million words, it could be anything. And the picture is functionally the same for a million and first word and so on.

If you want to find the steady-state distribution π, you can do so by solving the linear algebra equation

$$\pi = \pi T$$

along with the constraint

$$\sum_i \pi_i = 1$$

A conceptually simple (although potentially computationally expensive) way to do this approximately is to recall that

$$p_{i+m} = p_i T^m$$

In the limit that m is large, p_{i+m} will converge to π regardless of p_i. Since this is true regardless of p_i, it means that the rows of T^m must be approximately equal to each other and hence to π. So, if you multiply T by itself repeatedly until it converges, the rows of the resulting matrix will approximately equal π.

In an absorbing Markov chain, there is some state (possibly several) that always goes to itself with probability 1.0. If the frog lands on this lily pad, then he stays there forever. Absorbing Markov chains are used for modeling processes that terminate, such as the behavior of a visitor to a website. There could be a variety of pages they navigate through and maybe return to. But eventually, they will end up in the "made a purchase" state or the "left our site" state. Absorbing Markov chains can also be used to model the lifetime of a physical machine that eventually breaks down or a medical patient who eventually dies.

Sometimes, you will see absorbing Markov chains for describing situations that usually use irreducible chains. Natural language typically uses irreducible chains, but if you are trying to model short pieces of text such as e-mails or text messages, then you might want to add in an absorbing "the message is done" state.

With irreducible Markov chains, we tend to ask questions such as the following:

- What is the long-term equilibrium distribution π?
- Given the state that I'm in now, how many steps will it take before the probability distribution of my state approximates π?
- How many steps will it take, on average, for me to get from state A to state B?
- Given the state that I'm in now, what is the probability distribution for where I will be in 5 time steps?

Irreducible Markov chains are also building blocks in a range of other probabilistic models, which most of the rest of this chapter will be devoted to.

With an absorbing Markov chain, we are more likely to ask the following:

- Given where I am now, how long until I enter an absorbing state?
- Given where I am now, how likely am I to end up in each absorbing state?
- How many times can I expect to visit state A before I finally get absorbed?

25.3 Markov Chain Monte Carlo

In the previous chapter, we have actually already seen a fairly advanced application of Markov chains. The details of it are outside the scope of this book, but since we're talking about Markov chains in detail, I wanted to point out the connection.

Recall that we used a technique called "Markov chain Monte Carlo" (MCMC) to estimate the trained parameters of a Bayesian network given observations. Let me review what we did:

- We had a Bayesian network describing the relationships between several variables. Let's use θ to denote the parameters that characterize the network. The topology of the network was specified, but it was untrained, so θ was not known.
- We had prior distributions on θ, reflecting our initial confidences about what the "real" value of θ was.

- We had a dataset that we considered to be generated by that network.
- Taking the priors and the data into account, there is a natural way to "score" any particular set of parameters θ. It is the prior probability of θ times the probability of the data we saw given θ:

$$S(\theta) = \Pr(\theta)\Pr(Data \mid \theta)$$

- These scores do *not* define a probability distribution, since we have no guarantee that they sum up (or integrate) to 1.0. However, if we were to divide each score by the sum of all the scores, then we would have a probability distribution over all possible θ, indicating how likely (in the Bayesian sense) they are to be the real parameters.

Let's use π to denote this distribution over all the possible values of θ. Knowing π completely characterizes our beliefs and guesses about the correct value of θ. However, all we have as a guide to π is this potentially extremely complicated function $S(\theta)$. We know that $S(\theta)$ is proportional to $\Pr(\theta)$, but we don't know the constant of proportionality.

Knowing only the function $S(\theta)$, how are we to make inferences about π?

The most general solution to this problem is an MCMC simulation. It generates a sequence of samples θ_i, which form a Markov chain, whose steady-state distribution is equal to π. This means that it you generate enough samples, they will give you a representative sampling of π, and you can use them to estimate the average value of θ or any other statistic you desire.

There are a number of MCMC algorithms, and I won't get into their details here. Roughly though, to generate θ_{i+1} we start with θ_i and then add some random perturbation Δ and set $\Phi = \theta_i + \Delta$. Φ is a candidate value for θ_{i+1}. Using Score(θ) and Score(Φ), we probabilistically either accept Φ as the new θ_{i+1}, or we sample another Δ and try again. In this way, θ_{i+1} is a probabilistic function of θ_i but only θ_i.

If you use an MCMC algorithm for accepting/rejecting Φ, and sample Δ in a way that the every potential θ *can* be reached from every other, then you are guaranteed that the steady state of this Markov chain will be π. If you use a clever way to sample Δ, then it will, on average, converge to π quickly and the θ_i will be closer to independent samples.

25.4 Hidden Markov Models and the Viterbi Algorithm

One of the most important uses of Markov chains is as a building block in "hidden Markov models" or HMMs. Let me start with an example. Imagine you are reading through some blurry text, trying to figure out what it says. Let the correct words be denoted by random variables X_i, and let the blurry pictures be denoted by the random variables Y_i; your goal is to guess the sequence of the X_i given the known Y_i.

Your first thought might be to just guess that X_i is equal to whatever letter Y_i looks the most like, but that's not a perfect way to do it. Imagine that several of the words look clearly like "the duck said," but the next word looks ambiguous; it could be either "quick" or "quack," but it looks slightly more like "quick." In this case, you would probably say that the word is "quack," since it is vastly more likely given the context of talking about ducks.

To take another example, let's say that I have two coins: one is fair, and the other is heads 90% of the time. After each coin toss, I have a 10% chance of switching coins for the next toss and 90% of staying with the same coin. Let X_i denote whether the ith toss was using the biased coin and Y_i denote whether it came up heads. Can I identify the places where I switched from the fair to the weighted coin?

The situation where we have an underlying Markov chain X_i and observations Y_i that are dependent only on their associated X_i is called a "hidden Markov model" (HMM), and the X_i are called the "hidden states." The dependency structure of an HMM is often visualized as follows:

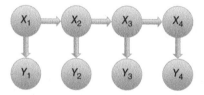

The HMM is characterized by

- The initial probability distribution of X_1
- The transition matrix of the X_i
- The conditional probability function $\Pr(Y|X)$.

There are a number of different analytical questions you can ask about Markov chains, but the most popular one is what is the most likely sequence of the X_i, given the observed Y_i. In our natural language example, knowing this sequence would give us a prediction of whether the ambiguous word is "quick" or "quack." When throwing fair and biased coins, knowing the sequence would let us see when we switched from one coin to the other.

I should note that finding the most likely sequence of X_i is not the same as finding the likeliest value for each of the X_i. To take a coin-flipping example, say that the following sequences have the following probabilities, and all other sequences have probability 0:

Sequence	Probability
HHT	0.4
HTT	0.3
HTH	0.3

In this case, the likeliest sequence is HHT with a probability of 40%, so the Viterbi algorithm will give us $X_2 = H$. However, X_2 actually has a 60% chance of being T. It's just that that 60% probability mass is spread out among several different sequences.

In the next section, I will discuss the Viterbi algorithm, which lets us find this optimal sequence. It's the most complicated algorithm that I describe in this book, so I wanted to devote a full section to it.

The times that HMMs really shine are when the observations are very ambiguous, but the X_i change only rarely. In these cases, even a human eyeballing the data often can't discern when the change happens, because the transitions are so subtle. I should note that in these cases, it is implicit in HMMs that the length of time X has a particular value is geometrically distributed. This is a very strong assumption, which often isn't satisfied. In practice, this usually doesn't end up being much of a problem, so long as the length of time at a particular X tends to be long.

25.5   The Viterbi Algorithm

The Viterbi algorithm is the easiest to understand if you look at the possible underlying states as what's called a "trellis diagram," showing all the possible hidden states and all the transitions between them. For concreteness, let's use the example of fair and biased coins. In that case, the trellis diagram looks like the following:

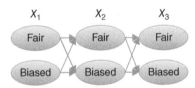

A sequence of hidden states corresponds to a path through this graph, from the first layer to the last. The probability associated with any particular path is

$$\text{Probability} = \left(\prod_i \Pr(X_{i+1} \mid X_i) \right) \left(\prod_i \Pr(Y_i \mid X_i) \right)$$

This would be a good point to back up for a second. No large-scale implementation of the Viterbi algorithm uses probabilities, because the probability of even the likeliest path is usually so small that the computer can't distinguish it from 0. So instead, we solve the equivalent problem of maximizing the log of this probability:

$$\text{Score} = \log(\text{Probability}) = \sum_i \log(\Pr(X_{i+1} \mid X_i)) + \sum_i \log(\Pr(Y_i \mid X_i))$$

Looking at it this way, you can see that a particular path's score is just the sum of the "scores" for all of it edges and nodes. Here the score of an edge is the log of its transition probability, and the score of a node is the log of the probability of the observation we made (assuming that we are in that hidden state).

Let's introduce two critical pieces of terminology:

- $P(i, x)$ is the highest-scoring path that goes up through the ith layer and ends at $X_i = x$.
- $S(i, x)$ is the score of $P(i, x)$.

That is, $P(i,x)$ is the best path through the trellis diagram *that ends at a particular node* in its ith layer. Given that we end up at (say) X-Fair on step 50, what is the highest-scoring path that would take us to that point?

The key insight behind the Viterbi algorithm (as applied to our case) is to find $P(i,\text{Fair})$ by conditioning on whether the $(i-1)$th state is Fair or Biased. We then see that either

$$P(i,\text{Fair}) = P(i-1,\text{Fair}) + [\text{Fair}]$$

or

$$P(i,\text{Fair}) = P(i-1,\text{Biased}) + [\text{Fair}]$$

whichever one has higher score. The score $S(i, \text{Fair})$ will then be given by

$$S(i,\text{Fair}) = \max \left\{ \begin{array}{l} S(i-1,\text{Fair}) + \log(\Pr(\text{Fair} \rightarrow \text{Fair})) \\ S(i-1,\text{Biased}) + \log(\Pr(\text{Fair} \rightarrow \text{Biased})) \end{array} \right\} + \log(\Pr(Y_i \mid \text{Fair}))$$

So, we have now reduced the problem of finding $P(i,\text{Fair})$ and $P(i,\text{Biased})$ to the problems of finding $P(i-1,\text{Fair})$ and $P(i-1,\text{Biased})$. If we can find those, then it is straightforward to find for i instead of $i-1$.

The Viterbi algorithm works by walking through the trellis diagram, filling in all of the $S(i,X)$ one layer at a time. Whenever it calculates $S(i,x)$, it keeps track of that node's "parent," that is, the node in the previous layer that led to it. When these have all been filled out, the pseudocode for the algorithm is then

```
Input:
        Observations Y₁, Y₂, ..., Yₙ
        Initial probabilities Pr(X₁)
Transition probabilities Pr(Xᵢ->Xⱼ)
Observation probabilities Pr(Y|X)
Initialization:
        Construct the trellis diagram
        For j=1...k:
                Node(1,j).score = Pr(Y₁|j)
```

```
Processing:
  For i=2…n
    For j=1…k
```
$$Node(i,j).parent = argmax_x \{Node(i-1,x).score + Ln(Pr(x|j))\}$$

$$Node(i,j).score = Node(i,j).parent.score + Ln(Pr(Y_i|j))$$

```
Constructing the output:
```
$$OutputStates(n) = argmax_x \{Node(n,x).score\}$$

```
  For i=n-1,…,1:
```
$$OutputStates(i) = OutputStates(i+1).parent$$
```
Output: OutputStates
```

25.6 Random Walks

One of the simplest stochastic process is called a random walk or sometimes a "drunkard's walk." The idea is that X_1 is some integer, usually 0. Then, in general,

$$X_{i+1} = \begin{cases} X_i +1 \text{ with probability } p \\ X_i -1 \text{ with probability } 1-p \end{cases}$$

The motivation that is given for this model is that you have somebody who is very drunk and doesn't know where they are going. At each point in time, they have a probability p of taking a step to the right and $1-p$ of stepping to the left.

If $p = 0.5$, the walk is called "unbiased." In this case, the walker will drift around through space aimlessly. In the long term, they will pass through the origin an infinite number of times. Their average location will always be at $X = 0$, but that's just because they are equally likely to be on the left or right. Their average distance from the origin will scale as the square root of the number of steps they've taken.

To be more precise, each step is a Bernoulli(p) random variable, so

$$X_{i+n} - X_i = 2 * \text{Binomial}(p,n) - n$$

25.7 Brownian Motion

Brownian motion is a generalization of random walks from the integers to the real numbers defined by

$$X_{i+1} - X_i = \text{Normal}(0,\sigma^2)$$

The only parameter to the model is σ, which measures how far it is likely to move in a single time step.

Historically, Brownian motion was the first stochastic process to be studied extensively. It was used to model the location of a particle suspended in a liquid, as it floats around aimlessly.

The most notable aspect of Brownian motion mathematically is that the difference between nonconsecutive locations is still a Brownian motion, just one with wider spread:

$$X_{i+t} - X_i = \text{Normal}\left(0, t * \sigma^2\right)$$

This is important for many applications. The motion of a particle, for instance, isn't really in discrete time steps. It moves continuously throughout time. But if we measure its location every millisecond, every second, or every hour, we will still see a Brownian motion.

At very small time scales, the motion of a particle is more like a random walk, as it gets jostled about by discrete collisions with water molecules. But virtually any random movement on small time scales will give rise to Brownian motion on large time scales, just so long as the motion is unbiased. This is because of the Central Limit Theorem, which tells us that the aggregate of many small, independent motions will be normally distributed.

Outside of physics, the most important application of Brownian motion is probably finance. When modeling the movement of a stock's price, we want to capture the idea that, absent any hunches or business developments, it will wander around aimlessly. This sounds like Brownian motion, but there is a catch; a 10% drop in price is equally significant no matter the starting price, and the price can never go negative. To achieve this "scale-free Brownian motion," we let X denote the logarithm of the security's price, rather than the price itself.

Is this an accurate model of how the prices of real stocks and bonds evolve? No, it isn't. However, it is a very common "null hypothesis" model, where we don't assume that there are any long-term trends, and we don't assume that there is any "correct" price that the motion tends to stay around.

25.8 ARIMA Models

For data science applications, the most important problem with Brownian motion is that it moves around without any sense of a "normal" value. Real processes tend to have a baseline that they hover around. The price of a security is a great example; it will fluctuate up and down, but not *too far* in either direction. This behavior is called "mean reverting," since you can think of it as an elastic force pulling the random variable back toward its average value.

The classical way to model this is called an "autoregressive moving average" or ARIMA. In general, an ARIMA model is defined by

$$X_{i+1} = c + \phi_0 X_i + \phi_1 X_{i-1} + \cdots + \phi_k X_{i-k} + \text{Normal}(0, \sigma^2)$$

However, you will typically just see this taken out to a single term:

$$X_{i+1} = c + \phi_0 X_i + \text{Normal}(0, \sigma^2)$$

where $0 < \varphi < 1$. In this case, there is a long-term average value of

$$E[X] = \frac{c}{1 - \varphi}$$

X will fluctuate around this average value, sometimes randomly moving fairly far from it. But it will always be pulled back to the mean.

25.9 Continuous-Time Markov Processes

Let's move back to Markov chains that have only a finite number of states – the frog hopping between lily pads. A limitation of these Markov chains is that they occur at discrete time steps. Sometimes, this granularity is appropriate, such as words in text. In other situations though, we are monitoring a system in real time, and it could change its state at any moment. For example, a server that is hosting a website could get a new visitor to handle at any point in time, and a visitor could leave at any point too. Basically, our amnesiac frog has no idea how long it has been since he arrived at his current lily pad, and he could hop away at any moment.

The best way to think of a continuous-time Markov process is that you have a collection of rates λ_{ij} from state i to state j. Imagine time to be broken up into very small moments of length Δ. Then, for each distinct i and j,

$$\Pr(X_{t+\Delta} = j \mid X_t = i) = \lambda_{ij} \Delta$$

The probability of staying in state i is just 1 minus the sum of all of these probabilities of changing. λ_{ij} is the rate at which probability mass flows from state I to state j, as a fraction of the mass in state i.

In calculus terms, if we let $p(t)$ denote the probability distribution over states at time t, then

$$\frac{d}{dt} p(t) = p(t) \Lambda$$

where Λ is a matrix of all the λ_{ij} (with 0s on the diagonal). We can find the steady-state distribution π by solving

$$\pi\Lambda = 0$$
$$\sum_i \pi_i = 1$$

similar to what we did with discrete-time Markov chains.

There is a critical difference between π for discrete and continuous chains though. For discrete Markov chains, π measured how often we transition into each state. In continuous processes, π measures the fraction of all time that is spent in each state. These are not the same thing, because it is possible for a lot of probability mass to flow into a particular state, but to then flow out very quickly. So, the system transitions into this state very often but only stays there briefly before moving on to a more long-lived state.

Continuous-time processes and discrete chains are both characterized by matrices, but those matrices are quite different. Both need all nonnegative entries, but the similarity stops there. For Markov chains, each row must sum to 1. For continuous-time processes, the sums are irrelevant, just so long as the diagonal of the matrix is always 0.

25.10 Poisson Processes

A Poisson process is used to model a stream of events that occur at random intervals. It is characterized by a single parameter λ, which is the average number of events per unit time. Alternatively, it is sometimes characterized by $\theta = 1/\lambda$, the average time between events.

There are a number of ways to think about a Poisson process, but in my mind, the easiest is this: break time up into many small intervals of size Δ. Each interval, independently of all the others, has probability $\lambda\Delta$ of having an event. In the limit of Δ being small, we converge to a Poisson process. It has the following properties:

- The interarrival times between consecutive events and exponentially distributed, with mean $= \theta$. This means that the time *can* be arbitrarily long but is weighted toward short time.
- Consecutive interarrival times are independent.
- For any time interval of length T, the number of events that occur in it is a Poisson distribution with mean λT. This means there *can* be arbitrarily many events, but huge outliers are not likely.
- If two intervals don't overlap, the number of events in them are independent of each other.

Poisson processes are a fantastic way to model many real-world systems. It is illustrative though to highlight several types of system that are not Poisson:

- One event tends to precipitate other events in rapid succession. An example might be trades that are made of a stock, where one person making a trade causes many other people to trade in reaction.
- An external force causes events to come in bursts. For example, visitors to a website might come in bursts because they all saw the same link that was posted.
- There is a tendency toward even spacing between events. If we replace a device whenever it wears out and breaks, we will probably not have to replace it again for a while because its parts are not worn out. This would mean that the time between events is not exponentially distributed.
- There is a small population of events that can happen, and they happen only once within a period. For example, an array of machines might fail at irregular intervals, and I fix any broken ones every morning. This means that if there are many failures early in the day, there will be fewer failures later in the day, because there are fewer machine running that could potentially fail.

The first two cases tend to be larger problems in practice, and their key property is that the different events are *not* independent of each other.

25.11   Further Reading

Harchol-Balter, M, *Performance Modeling and Design of Computer Systems: Queueing Theory in Action*, 2013, Cambridge University Press, Cambridge, UK.
Ross, S, *Introduction to Probability Models*, 9th edn, 2006, Academic Press, Waltham, MA.
Feller, W, *An Introduction to Probability Theory and its Applications*, Vol. 1, 3rd edn, 1968, Wiley, Hoboken, NJ.

25.12   Glossary

Absorbing state A state in a Markov chain that goes only to itself.
Absorbing Markov chain A Markov chain with at least one absorbing state.
Ergodic Markov chain A Markov chain in which every state can be reached from any other.
Equilibrium distribution A probability distribution over the states of a Markov chain that stays the same as the chain evolves by one time step.
Markov property The key assumption for Markov chain. A state in a Markov chain can depend probabilistically on the state right before it, but only the one right before it. If you condition on knowing X_i, then X_{i+1} is independent of all X_j with $j < i$.

Markov chain A collection of states and transition probabilities between them, where each state depends only on the one before. Many Markov chains also require a probability distribution over the starting state.

Poisson process A way to model sequences of events that happen at random intervals. The times between consecutive events are i.i.d and exponentially distributed.

Stationary distribution Synonym for equilibrium distribution.

Transition matrix A matrix specifying the probability of transitioning from every state to every other in a discrete Markov chain.

Parting Words: Your Future as a Data Scientist

So you've read this book, and let's assume for the sake of argument that you thought the subject matter was pretty cool (and that the writing was brilliant, of course). What now?

My most important advice is to get out there and start tackling some real problems. I've done work as a software engineer and an academic, and I'm constantly impressed by how much more intellectually dynamic data science is than anything else I've done. In a single day, I will flit between low-level debugging, designing software architecture, helping clients to translate a business problem into math, and brushing up on my linear algebra. In data science, there is always something new that you can learn, and usually something new that you *have to* learn, and no book can substitute for real experience in that kind of environment.

As far as broadening your knowledge base, there are several directions (not mutually exclusive) that you might consider growing:

- Really the best, if you have a particular area of application in mind, is to become more of a domain expert in whatever it is you want to apply data science to. Remember that the key to doing great data science is to ask the right questions, and the only way to do this is to have a deep understanding of the domain you're studying.
- A lot of data scientists almost double as software engineers. I would probably put myself in this category. They know a number of additional programming languages, they've written multithousand-line pieces of code with many interacting parts, and they are well versed in their computer science fundamentals. This is a great direction to go if you want to work in a start-up-type environment or in Big Data.
- Some people get deeply immersed in machine learning. This will serve you well if you want work that is more academic in flavor or if you specialize in solving a few extraordinarily hard problems at a large company.
- Some data scientists grow in the direction of statistics, learning more about A/B testing, how to design experiments, and all the various things you can do

The Data Science Handbook, First Edition. Field Cady.
© 2017 John Wiley & Sons, Inc. Published 2017 by John Wiley & Sons, Inc.

to eke insights out of small datasets. This approach is more common in large companies, where people have the ability to specialize.

The range of datasets available to study is changing quickly, but I want to point out a few trends that will probably be relevant to you:

- Data science to date has been dominated by data from the Internet (web logs, click-throughs on ads, etc.), but the emphasis is shifting toward data from the physical world.
- Smartphones produce a lot of cool data and will continue to play a more central role in data analysis. In particular, they measure a number of different things at once (such as location, motion, and whatever you are looking at on your phone), and very few people have done much to bring these different data streams together in a unified analysis.
- The "Internet of things" refers to the idea that sensors are becoming ubiquitous, and many devices that are neither computers nor cell phones will be producing data that can be accessed through the cloud. This means sensor nets, machines in factories, toasters – you name it.
- The "quantified self" movement refers to the fact that there is more and more measurement of one's own biological stats. This means everything from heart rate, to blood glucose, to breathing rate. As wearable sensors become cheaper and more ubiquitous, more of this data is becoming fine-grained time series.

The software and hardware tools we use are also evolving. For example:

- I didn't have a chance to get to this in the book, but graphics processing units (GPUs) are specialized hardware that include many processing cores for massively parallel processing. They aren't always helpful, but they can be extraordinarily powerful in many cases. Deep learning is a good example of an area that really benefits from them.
- People will probably realize that cluster computing is very overhyped right now, and some of the wind will probably get lost from the Big Data sails.
- Don't get me wrong though: Big Data is still going to remain a huge deal! For many datasets, a cluster will make the analysis much quicker; in other cases, it will be the only way to get the job done. The big difference is that people will be better at identifying when a cluster is or isn't the right tool for the job. I also expect that Big Data will become much more seamlessly integrated with local machines and that it will often be invisible to the user whether their computation is running on their local machine or a cluster in the cloud.

Data science in the coming years is going to be a very exciting journey. Hopefully, this book has gotten you off on the right foot.

Best regards,
Field Cady

Index

The Data Science Handbook, First Edition. Field Cady.
© 2017 John Wiley & Sons, Inc. Published 2017 by John Wiley & Sons, Inc.

Printed and bound by CPI Group (UK) Ltd, Croydon, CR0 4YY

27/10/2024

14580473-0001